# The Homeowner's DIY Guide to Electrical Wiring

D0879476

# The Homeowner's DIY Guide to Electrical Wiring

David Herres

Mc
Graw
Hill
Education

New York   Chicago   San Francisco
Athens   London   Madrid
Mexico City   Milan   New Delhi
Singapore   Sydney   Toronto

McGraw-Hill Education books are available at special quantity discounts to use as premiums and sales promotions or for use in corporate training programs. To contact a representative, please visit the Contact Us page at www.mhprofessional.com.

**The Homeowner's DIY Guide to Electrical Wiring**

1 2 3 4 5 6 7 8 9 0   DOC/DOC   1 2 0 9 8 7 6 5 4

ISBN    978-0-07-184475-8
MHID    0-07-184475-9

This book is printed on acid-free paper.

| **Sponsoring Editor** | **Project Manager** | **Indexer** |
|---|---|---|
| Roger Stewart | Patricia Wallenburg, TypeWriting | Claire Splan |
| **Editing Supervisor** | **Copy Editor** | **Art Director, Cover** |
| Stephen M. Smith | James K. Madru | Jeff Weeks |
| **Production Supervisor** | **Proofreader** | **Composition** |
| Pamela A. Pelton | Claire Splan | TypeWriting |
| **Acquisitions Coordinator** | | |
| Amy Stonebraker | | |

## About the Author

**David Herres**, Master Electrician, is the owner and operator of a residential and commercial construction company. He is the author of *2011 National Electrical Code® Chapter-by-Chapter*, *Troubleshooting and Repairing Commercial Electrical Equipment*, and *The Electrician's Trade Demystified*, all published by McGraw-Hill Education. Mr. Herres has written more than 150 articles for construction and electronics magazines, including *Electrical Construction and Maintenance*, *Cabling Business*, *Nuts and Volts*, *Solar Connection*, *Fine Homebuilding*, and others.

# Contents

# Acknowledgments

I wish to thank all those at McGraw-Hill Education whose diligent work in producing this book helped to make it a reality. Thanks to Roger Stewart, my editor in San Francisco, who was with the project from start to finish, and to Judy Bass, in New York, who provided initial inspiration and guidance. Their colleague, Amy Stonebraker, brings new meaning to the word *competence*.

*David Herres*

# Introduction:
# Why Do Your
# Own Wiring?

For a long time, the best advice seemed to be "hire a professional and get it done right." Depending on the task contemplated and the individual's level of expertise, this notion may or may not be valid. You have to assess the situation and decide what will work for you. Even if you hire out all the work, electrical knowledge and expertise, as conveyed in this book, will be of value in completing your building project and maintaining it in the future.

## The Homeowner's Role Is Expanding

It is a fact that nowadays homeowners are far more active in their construction projects, whether new building or remodeling. If professionals are hired, the owners still may play a prominent role in planning and moving forward with the job. Often the homeowner is the builder. In some cases, though, a professional is hired as a designer-advisor and, in jurisdictions where there is oversight, interfaces with the inspector and, if necessary, signs off on the job.

All of this is especially true in the area of electrical work. Some home crafters draw a line in the sawdust, avoiding electrical work altogether. Others set the

National Fire Protection Association, NFPA, National Electrical Code, NEC, NFPA 70, and NFPA 70E are registered trademarks of the National Fire Protection Association. All NFPA trademarks are the official property of the National Fire Protection Association.

boxes, drill studs, and pull cable, leaving all terminations to the electrician. The extreme case is to do everything including the service without hiring a professional at all. This approach will maximize the monetary savings for the homeowner, and of course there is immense satisfaction in doing it all. If this is too great a leap into the unknown at this time, it might be something to aim for down the road.

Electrical work is very exacting and presupposes accurate knowledge (on an open-book basis) of the *National Electrical Code*® (*NEC*®). This document is applicable in the United States, Mexico, Venezuela, and certain other countries. In Canada, it is the *Canadian Electrical Code* (*CEC*), and in Europe it is the *International Electrotechnical Commission* (*IEC*). Australia and New Zealand recognize the *Australian/New Zealand Standard for Wiring Rules*. These codes are similar, differing for the most part in only a few details here and there. In this book, we will be referring to the *NEC*. If you are in an area where another code applies, it will be a question of referring to your documentation and making the necessary adjustments.

## Why Get Involved?

Homeowners choose to do their own electrical work for a variety of reasons:

- **To save money.** In the building trades, electricians are among the most highly paid in terms of hourly rate. No matter how fast the individual works, electrical installation is time-consuming. In 2014, expect to pay over $4,000 to have a new small residence wired, including the service but without extensive data networking or home automation and not including light fixtures or appliances.
- **To impress family members, neighbors, and colleagues at work.** This is where preparation really pays off. The *NEC* stresses throughout that electrical equipment is to be installed in a "neat and workmanlike manner." A large portion of the job will be concealed behind building finish surfaces and sometimes underground, but the work in progress will be watched carefully by those in the area. A portion of the work will remain visible for the life of the building, and onlookers will be impressed if it is a first-rate job. Thus, beyond the issues of safety and efficiency, there is a great need to produce an outstanding product, and here again, knowledge and expertise are decisive factors in meeting this goal.
- **To increase self-esteem.** Self-esteem is an important motivation for the home crafter-electrician. We all like to consider ourselves good at what we

do, and with research and practice, the general trend is to improve. Every project you undertake and bring to a successful conclusion will contribute to your ability in the future to tackle a more complex or difficult task, and for many of us, this goes way beyond the dollars saved.

- **To be in touch with some fundamental processes of the universe.** It is a palpable pleasure to channel electrons through conductors and watch the way these elementary particles react when we throw the switches. Completing an electrical wiring project puts us in touch with some fundamental processes of the universe, and there is a great deal to be said for that.

## Applicable Mandates

Nonelectrical work, both residential and commercial, is governed by multiple building codes, and most of these are less restrictive and detailed than the *NEC*. The *Building Officials' and Code Administrators' (BOCA) Plumbing Code*, for example, lays out general principles such as those intended to ensure that drain water will not infiltrate the drinking water system, but it is a comparatively slim volume, and the requirements are less detailed and specific than the *NEC*.

Why all this oversight of electrical work? In a nutshell, it is to protect end users from the twin demons of fire and electrical shock. The *NEC* has had great success in this regard. In recent years, the number of nonutility electrical shock fatalities has gone way down. This decline has been due largely to the increasingly broad *NEC* mandate requiring ground-fault circuit interrupters (GFCIs) in more locations. Homeowners and builders may gripe at the initial cost of installation and instances of nuisance tripping, but these are small prices to pay when you consider that in the fullness of time, little fingers will seek ways to insert metal objects into receptacles and impatient construction workers will saw off the ground prongs of power tools that may be used in wet environments.

Similarly, though in an earlier stage of development, the arc-fault circuit interrupter, where installed, is a highly effective guard against electrical fires, although here again there is the issue of nuisance tripping. In Chapter 1, we'll talk about these lifesaving devices in greater detail—theory of operation, where required, where prohibited.

Electrical codes are an essential part of the picture but not the whole story. They are intended to provide protection from the hazards that can arise in connection with the use of electricity. The savvy home electrician needs other intellectual tools as well. A working knowledge of high school math is essential. Don't worry about calculus or advanced trigonometry, but you will need to perform simple operations such as solving for an unknown in linear algebraic equations and finding a square root with the aid of a hand-held calculator.

As we have indicated, electrical work is a big subject, but for the homeowner, where the field is limited to residential construction, it is a bit simpler. To conclude your project, you'll have to adopt a methodical, step-by-step approach. There's good news, though! It's all open-book, meaning that when a question arises, you can consult the *NEC*, the Internet, electronic textbooks, and this book to find the answers you need.

Herein we begin with some basics and proceed into more difficult areas the home crafter-electrician is likely to encounter. If you are in an early stage in this interesting and rewarding undertaking, start at the beginning, and you will not have a problem tackling common electrical jobs in a residential setting. More advanced readers can jump around, filling in bits of knowledge here and there with a goal of seeing the picture in its entirety.

# The Homeowner's DIY Guide to Electrical Wiring

# CHAPTER 1

# Avoid Building Fire and Shock Hazards into Your Work

A
s everyone knows, there are hazards inherent in the use of electricity. In residential work, they are less intense than in a factory or commercial setting, where the voltage levels and available short-circuit currents are much higher. Notwithstanding, great care is also needed in home wiring. You should remember the child who is sleeping upstairs or playing outside in a wet area near a receptacle or appliance you wired. Proper design and installation procedures will protect against fire and shock hazards and prevent tragedies from occurring. We'll examine the hazards and see how they can be mitigated.

## How Electrical Fires Start

More fatalities result from electrical fires than shock, and most of them are caused by smoke inhalation. In residential work, you must keep in mind where the greatest dangers lie and take proactive measures to guard against them.

Starting with the greater potential hazard, how does an electrical fire begin? The major culprits are series and parallel arc faults and conductors that overheat due to insufficient ampacity. The first of these is an installation deficiency, and the second is more likely a design miscalculation. Either can result in a fiery inferno with great property damage or, far worse, injury or loss of human life.

Arcing faults, series or parallel, can initiate a fire. Series arc faults do not usually increase the load (except sometimes in the case of a motor), as seen by the overcurrent device, fuse, or circuit breaker. Therefore, they do not cause the over-current device, fuse, or circuit breaker to trip out. The fault continues until either it clears itself by burning out the connection and breaking the circuit's continuity or nearby combustible material is ignited, perhaps destroying the entire building. This can happen even where the fault is inside a metal enclosure, although such enclosures usually limit the risk.

Series arc faults usually result from one of these causes:

- An errant nail or drywall screw partially penetrates a conductor, not sever-ing it or shorting it out so as to interrupt the circuit, but reducing the current-carrying capacity and making a local hot spot.
- A termination is improperly torqued. A medium-sized residential job will involve thousands of terminations in the branch circuits, not to mention the service, which also can be problematic. The usual faults are a screw terminal in a breaker, entrance panel, or load center or a switch, light fix-ture, or receptacle that is not sufficiently tight (or too tight) and a wire nut that is not tight enough, contains misaligned conductors, or is disrupted when stuffed back into the box.

## Avoiding Arc Faults

Some steps can be taken to avoid the problem of nail or screw penetration. First and foremost, you should locate drilled holes in studs and joists near the center of the framing member so that there is less chance of the fastener reaching or pene-trating the wire through floor, ceiling, or inside or outside wall finish. Also, 2 × 6 studs provide more isolation than 2 × 4s. If the cable segments between adjacent studs are not pulled too tight, there is a better chance that the cable will deflect when found by a nail or screw. In this connection, there is a greater chance of damage when air nailing is done because the fast-moving projectile will penetrate the cable rather than push it forward. The *National Electrical Code*® (*NEC*®) addresses this situation by requiring all conductors to be 1¼ inches from the outer edges of the framing unless protected by metal deflection plates made for the pur-pose, as shown in Figure 1-1. The bottom line is that when you install wiring that is not in metal raceways, this scenario always should be kept in mind.

Another defect in a residential wiring job that may cause dangerous electrical arcing is an improperly torqued termination, either too loose or too tight. If you coil the wire under the screw terminal of a circuit breaker, switch, receptacle, or light fixture, neglecting to tighten it sufficiently, there will be poor metal-to-metal

**FIGURE 1-1** This metal plate protects cable closer than 1¼ inches from the edge of a framing member.

contact. If a significant amount of current must flow through this bottleneck as a result of the applied voltage and the connected load, the electrons will seek an additional path and end up traversing an ionized air gap, forming miniature lightning bolts that emit all kinds of radiation, including light and heat, with a characteristic frying sound. Over a period of time, the heat will further degrade the electrical joint, increasing the resistance, making still more heat, and so on. Eventually, the fault will burn clear, and the arc will be extinguished or the amount of heat will be great enough to ignite nearby combustible material, metal enclosure notwithstanding (Figure 1-2).

The same thing can happen when a wire nut is not screwed tight or when the wire ends are misaligned. Frayed extension cords, electric motors with worn brushes, old light fixtures, and many other types of electrical equipment can exhibit dangerous arcing, and this is a frequent cause of electrical fires.

Electrical terminations can be overtightened as well. The threads may be stripped or stressed to the point where the connection fails some time in the future. You should have a good sense of how tight to turn the screw terminals in residential switches, receptacles, circuit breakers, and the like. For larger electrical equipment, it is necessary to use a torque screwdriver or torque wrench, following the manufacturer's specifications in the installation manual.

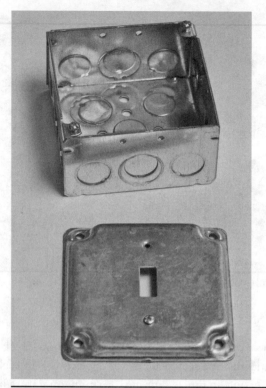

**FIGURE 1-2**   A metal enclosure usually contains heat from an arc fault so that the building will not catch fire—but not always.

## Arc-Fault Circuit Interrupter

A new technology has emerged in recent years. It is the *arc-fault circuit interrupter* (AFCI), which is capable of detecting an arc fault within a live circuit and interrupting the current flow before there is sufficient heat to initiate a fire. The device frequently takes the form of a circuit breaker. Installed in an entrance panel or load center, it detects any arc fault in the connected branch circuit or load and responds by tripping out. It is also sensitive to overcurrent like a conventional fuse or circuit breaker and performs that function as well. How does it work?

Internal solid-state circuitry monitors the current passing through the device. An electric arc, because of the rapidly fluctuating, harmonic-rich, irregularly intermittent, spiky nature of the waveform, is detected by the AFCI in accordance with internal algorithms. The device is a switch that opens the circuit. It will not reset or operate continuously until the defect has been located and repaired. The initial cost of installation is a small price to pay when you consider the AFCI's great potential for saving property and lives.

A disadvantage is that under some conditions, an AFCI may engage in nuisance tripping. In such a situation, you should resist the temptation to replace the AFCI with a conventional circuit breaker because then, although the lights will be back on, the arc fault may lie dormant for a period of time and then deteriorate further and cause a full-scale electrical fire.

Generally, AFCIs perform quite well. They should be used for circuits that supply power to all living spaces, not just bedrooms, as required by earlier codes. Living space in this context includes living rooms, dining rooms, hallways, and closets. In addition, AFCIs may be deployed for extra protection in areas where they are not required.

## Removing Accessible Abandoned Wiring

Abandoned wiring, no longer supplying power to a load and not tagged for future use, if it is accessible, should be removed. Because it is not energized, it cannot directly cause a fire, but if it is ignited by some other fire, even a nonelectrical fire, it can add dramatically to the fire load. Most burning electrical insulation produces a thick, choking smoke that displaces oxygen in the air and may cause severe injury or worse. If the abandoned wire is allowed to accumulate over time, it can become a burdensome liability to the building owner, and its presence makes troubleshooting faulty wiring more difficult.

These are the most common fire hazards associated with electrical wiring. There are others as well. Poor grounding sets the stage for lightning to find its way into the building. Additionally, faulty appliances, ranging from automatic washers and dryers to TVs and computers, even laptops, can burst into flames without warning. We will be discussing problems of this sort in greater detail later on.

A well-designed fire alarm system is essential. This may consist of simple smoke detectors powered by 9-volt batteries and wired into the alternating-current (ac) system to provide redundant power, as shown in Figure 1-3. Additionally, they may be wired together to sound in concert if smoke or heat is detected.

A definite upgrade, sometimes seen in large upscale homes, is a full-scale supervised fire alarm system, as currently installed in most nonresidential occupancies. Such a system is very expensive—a complete system in a small hotel can cost $100,000 or more. However, such systems offer very robust protection with automatic notification of the fire department or monitoring agency through two

**FIGURE 1-3**   A smoke detector powered by a 9-volt battery and an ac branch circuit.

dedicated redundant telephone lines that are automatically verified based on a predetermined schedule.

By *supervised*, we do not mean that there is a human seated at a console watching a bank of monitors at all times. The supervisory function is electronic, automatic, and continuous. If the integrity of any part of the system, including the control panel, becomes problematic, a trouble alarm will sound, and details will be shown on an alphanumeric display so that the defect can be located and corrected. We'll have a lot more to say about this upscale option in Chapter 12. For now, the point is that such a system affords excellent protection from electrical (or any other type of) fire in the home and that it is an option to be considered if budget permits.

In a similar vein, there is talk of requiring sprinkler systems in all new residential construction, and this may become a reality within a very few years. A more economical option would be to install sprinkler heads under the ceiling in each room, with dry piping to an outside fire hose connection that can be pressurized by firefighters using a tanker truck or hose run from a hydrant, obviating the need for a high-pressure, high-volume water supply to the home with a sprinkler system valve body and all that entails. As in a commercial or industrial system, only heads above the hot spots would melt out so that maximum water would be directed where needed, and there would not be extensive water damage in unaffected parts of the building.

## Electric Shock

The other of the two hazards in electrical installations is shock. Although statistically less prevalent than electrical fire fatalities, those caused by electric shock are gruesome in the extreme. Children, with active minds and inquisitive fingers, find ways to make contact with lethal voltages. These also can strike unsuspecting adults who handle poorly grounded power tools or frayed wires. Such accidents are preventable. It is the responsibility of the home crafter-electrician as well as the professional to build shock-proof installations insofar as possible. This includes, when doing repairs, additions, or retrofits to existing wiring, inspecting the overall system and making sure that it is safe. The most basic aspect of a wiring system from a safety-from-shock point of view is adequate, reliable bonding and grounding.

Most electrical systems are grounded. There are some specialized types of systems that are permitted to be ungrounded, which is to say that neither of the two conductors that are connected to the electrical supply is also connected to the earth so as to be at ground potential. Both sides of the circuit float above ground potential so to speak. (These ungrounded systems are not seen in residential occupancies.) Even where the electrical system is ungrounded, a grounding conductor is to be connected to earth with a full-scale grounding electrode system so as to be at ground potential. It is to be run along with the circuit conductors throughout the premises wiring and to be connected to all metal enclosures such as junction boxes, wall boxes, metal light fixture housings, and so on. The purpose in grounding these conductive objects is so that if, because of a fault such as a chafing wire inside a power tool or light fixture, the metal casing were to become energized, the full available fault current would rush through the entire input end of the circuit, including the fuse or circuit breaker. This would instantly trip out, interrupting the circuit and deenergizing the faulted metal enclosure.

---

### Grounding versus Bonding

Grounding and bonding are two separate but related concepts. *Grounding* refers to connection of a wire or circuit to the earth for the purpose of setting it at ground potential. *Bonding*, in contrast, is an intentional connection of two or more conductive bodies together or to the electrical system neutral for the purpose of keeping them at the same potential. Many inexperienced workers throw in an extra ground rod and believe that they are doing something great, whereas bonding back to the neutral bar is often far more effective.

## Grounding and the Breaker

Without the grounding conductor, the breaker would not trip. It is correct to say that the grounding conductor facilitates operation of the overcurrent device. The enclosure would remain energized until touched by a person who is also in contact with the ground (as is usually the case), and the individual would experience an electric shock. Its severity would depend on the level at which the enclosure was energized, the nature of the victim's contact with ground, the electrical resistance of the individual's body, the route the electric current took through the body, the duration of the shock, and other factors.

As we have seen, the equipment-grounding conductor in conjunction with the overcurrent device is a wonderful safety feature, but of course, it won't work if it is not present or if there is a break in the continuity anywhere along the line. We have all seen instances where the ground prong of an extension cord or power cord has been sawed off to make the plug fit into an old two-wire receptacle. The person who does this is either incredibly ignorant or guilty of depraved indifference to human life.

In the preceding discussion, we mentioned both grounded and grounding conductors. They are both at earth potential, and they are connected to the neutral bar within the entrance panel, but they serve different purposes, and they are color-coded differently.

The grounded conductor is the return or neutral side of the circuit that powers the load, and it carries the full amount of current that passes through the load. The insulation is white. The grounding conductor is that third wire that under normal nonfault conditions does not carry current and is not part of the circuit that powers the load. The wire is usually bare or has green insulation.

The grounded and grounding conductors are at the same potential because they are connected together in the entrance panel by means of the main bonding jumper. This is to be the one and only connection between the grounded and grounding conductors. Additional subsequent connections between them anywhere along the line, including within the load, are prohibited and would result in dangerous circulating currents. To emphasize, the grounded and grounding conductors are solidly connected together within the entrance panel, never to rejoin.

## Testing the Installation

It is part of the job of the home crafter-electrician or professional when doing any additions or modifications to premises wiring to review the existing system and ensure that this grounding conductor system is in place, intact, and has impeccable continuity. Fortunately, there is an inexpensive tester, a *circuit analyzer*, that permits the user to quickly test all the receptacles in the home to verify that they are wired properly. It plugs into the receptacle to be tested. There are three indicator lights, and referring to the key printed on the plastic housing, you can easily determine the status of the wiring. Lighting patterns differ, but on a typical model, two outer lights indicate correct wiring, no lights mean that the power is out, and other patterns mean that the hot and neutral are reversed or that there is some other fault in the wiring.

Besides verifying equipment-grounding conductor continuity and correct connection at all receptacles, there is another important test that can be performed on electrical loads to determine that the equipment ground is in place. The test instrument is the very inexpensive *neon test light*, a small plastic housing containing a single neon bulb, as shown in Figure 1-4. There are two short leads that are test

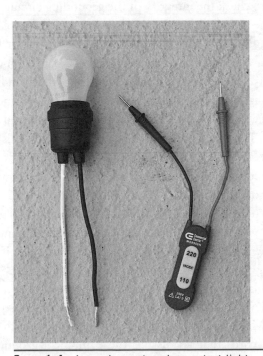

FIGURE 1-4    Incandescent and neon test lights.

probes, and this simple tester will yield a lot of information about the status of an electrical circuit and connected load.

---

CAUTION  *A neon test light should never be used where there could be more than 600 volts because then you are getting a little too close for comfort. High voltages can ionize a channel through the air, suddenly arcing to your body, seeking a path to ground.*

---

The neon test light can be used to check for the presence or absence of voltage, where an accurate measurement is not needed. If you touch the probes (either way, the polarity does not matter) to the terminals or wires in question, the neon bulb will light if energized, and the brightness will give a rough indication of the voltage. It is easy to tell the difference between 120 and 240 volts. At 90 volts, the bulb will glow faintly. A good way to get a sense of this is to try the test light on different voltages that are available in an entrance panel. You can also distinguish direct current (dc), which lights just one side of the bulb.

This little tester is great for troubleshooting an entrance panel when there is an outage. You can see whether either or both of the legs are live with respect to the neutral bar and whether there is voltage at the input of the main breaker and at the output of individual branch circuit and feeder breakers.

You can use the neon tester to check the integrity of the grounding circuit. The thing to remember is that the neon bulb, along with a resister inside the plastic package that is in series with it, draws a minute amount of current when it fires up. If you touch one probe to a live terminal and leave the other probe unconnected, the bulb will glow dimly. This happens because the free air around the unconnected probe has some slight ground potential. If you touch the unconnected probe with your finger, the bulb will glow considerably brighter depending on how well grounded you are. It is recommended that you do not do this old electrician's test unless you have some way to verify that the tester does not draw too much current and has not acquired an internal short and that you know for certain that the hot terminal is not at over 150 volts to ground.

If you touch the other probe to the neutral (white) conductor, the bulb will glow at maximum brilliance for that voltage. If you touch it to an intact equipment-grounding wire or terminal, it will glow at the same level. Therefore, this is the way that you can verify the integrity of an equipment-grounding conductor.

There is another test that this little instrument performs very well. You can find out whether there is a low-level or full-scale fault that is energizing an equipment or light fixture housing. Make up a test cord (properly labeled so that it won't be used as an extension cord) that is missing the equipment-grounding conductor. Through the test cord, power up the equipment, and touch one probe to an unpainted portion of the metal cabinet and the other probe to a known good ground. If there is light, you've got a problem.

## Ground-Fault Circuit Interrupters

There is an additional level of protection that has been very successful since its use became widespread in the 1960s. It is the *ground-fault circuit interrupter* (GFCI). With increased usage in the decades that followed, nonutility fatalities from electric shock have declined dramatically, particularly in the home. This lifesaving device does for shock prevention what the AFCI does for the prevention of electrical fires.

The GFCI can take the form of a circuit breaker for use in an entrance panel or load center, a small plastic-enclosed device molded into the power cord close to the plug of certain power tools and consumer appliances that may be used in wet areas, and a receptacle for use in kitchens, bathrooms, basements, outdoors, and in other potentially wet locations where individuals may become solidly grounded and thus subject to severe shock. The GFCI works by measuring the amount of incoming current on the ungrounded (hot, black) conductor coming from the entrance panel or load center and comparing it with the amount of current moving through the grounded (neutral, white) conductor going back to the supply. Under ordinary, nonfault conditions, these two amounts of current are the same. The GFCI interrupts the circuit if it detects a difference between 4 and 6 mA. [A milliamp (mA) is 1/1,000 of an amp.] Specifically, it is functioning correctly if it does not trip out with a difference of less than 4 mA and it does trip out with a difference of over 6 mA. This simple device saves many lives every year.

The receptacle-type GFCI is less costly than the breaker type, and it is commonly used in specified locations within all new residential construction (Figure 1-5). It is as easy to wire in place as a conventional receptacle except that it is a little more bulky and may present a problem if the wall box also contains wire nuts. You need to plan ahead by installing deep wall boxes wherever GFCIs will be installed.

The receptacle-type GFCI has a great advantage, which is that it can be used to power conventional downstream receptacles and in so doing extend GFCI coverage to them. You will notice on the back of the device a pair of terminals labeled "line" and another pair of terminals labeled "load." If you daisy-chain your downstream receptacles from the load terminals, they become, in effect, GFCIs. The box in which a GFCI comes packed contains stickers saying, "GFCI PROTECTED,"

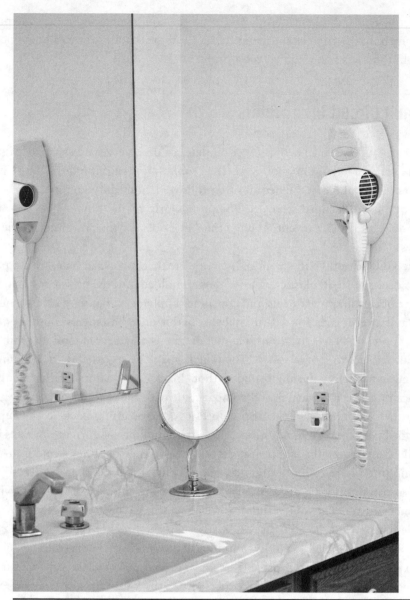

FIGURE 1-5    A hair dryer with a cord-type GFCI is plugged into a receptacle-type GFCI.

and they should be applied to any downstream conventional receptacles so protected. They are Code compliant wherever GFCIs are required.

Receptacle-type GFCIs have another important application. Existing homes are sometimes encumbered with obsolete two-wire receptacles supplied by old two-wire cable that does not have the third equipment-grounding conductor.

Often a decision is made to replace these obsolete receptacles with new equipment-grounding receptacles. The problem, however, is that it is not practical to run cable back to the entrance panel because finish wall and ceiling material are in place. If it is possible to stub a short length of Wiremold through a drilled hole and up from the basement, that is the way to go. In no case should an improvised grounding conductor be run to a nearby radiator or outside to an isolated floating ground rod. One solution that is Code compliant is to replace the two-wire receptacle with a GFCI. These devices do not depend on an equipment-grounding conductor to operate. When this substitution is made, use the stickers that say "NO EQUIPMENT GROUND" on the GFCI and all daisy-chained conventional receptacles.

GFCIs, because they are sensitive to very small amounts of fault current, may be subject to nuisance tripping, and because of this, you may lose a whole string of downstream daisy-chained receptacles. Typically, the reset button will not stay down when it is depressed. This will happen if there is a downstream ground fault, if the device itself has gone bad, or if the GFCI is not receiving power. The newer GFCIs incorporate light-emitting diodes (LEDs) that light up when the device trips out, but of course, the device won't light if there is no incoming power, so this makes it possible to partially troubleshoot the circuit without even taking off a wall plate.

The first step is to unplug all equipment connected to the GFCI or any downstream conventional receptacles that are GFCI protected. Often this clears the fault, and it is found that a toaster, lamp, or similar appliance is the culprit. If this does not provide an answer, and if the GFCI does not have a LED, pull the GFCI out of the wall box and see if there is power on the line terminals. (The neon tester is perfect for this.) If there is no power on the line terminals, there is a branch-circuit problem, and it is necessary to work back toward the entrance panel. If there is power on the line terminals, while you have the GFCI out of the wall box, disconnect the downstream string. You can do this by disconnecting just the hot wire (black or other color that is not white or green) from the load terminal. If this restores operation, you know that the problem is in the downstream string. Go to an easily accessible receptacle near the middle of the run, and take a reading. You will now know which half of the run contains the fault. Continue dividing the faulted segment in half and testing until you have isolated the fault. It may be a bad receptacle or, if the wiring is old, a chafing wire at the connector. It doesn't take a full-scale short that would trip the overcurrent device, just a slight leakage, to make the GFCI break the circuit. If this is the case, it may be possible to loosen the connector and repair the fault using electrical tape. If this doesn't work, you will have to remove some wall paneling and rewire.

This is the basic procedure for finding the fault when a GFCI refuses to reset. It is generally successful unless the fault is intermittent. Does it occur only when there is a driving rain or high humidity? Perhaps there is a leak in the roof, and water is coming down inside the wall. Sometimes it is possible to find the fault by applying water to the outside wall by means of a hose. This is especially true if there is an outside receptacle that is bugged off an indoor GFCI.

## Other Safety Issues

Electrical safety is always a work in progress. On the job site, be ever vigilant. The handles of insulated tools should be inspected periodically for any sign of deterioration. Minute punctures or cracks can hold conductive grease and moisture. Above all, ground the work, not the worker.

In a damp location such as a trench, use cordless tools when possible. Otherwise, make sure that your power is GFCI protected and that the equipment-grounding conductor has continuity back to the service.

CHAPTER **2**

# Basic Rules for Impressing the Electrical Inspector

Electrical installations are regulated by the local jurisdiction. In most cases in the United States, the states oversee the licensing of electricians. A construction permit is issued by the county or municipality, and it must be in place before work may begin. On completion, the owner or electrical contractor calls in for an inspection. The inspector visits the site and examines the work to see if the job corresponds to plans filed with the jurisdiction and to ascertain that the installation complies with the *National Electrical Code* (*NEC*) and any other standards that the state or local jurisdiction may have enacted. The inspection may be cursory and pro forma, but more often the inspector examines the work in detail and red tags anything that is not just right. If the work complies with applicable standards and the scope of the work does not exceed or fall outside the terms of the original construction permit, the inspector issues an operational or occupancy approval, with or without conditions, such as minor adjustments that must be made or limitations on future use.

## Mitigating Hazards

The key element in all of this is the *NEC*. This is a thick volume of requirements and mandates that cover every aspect of residential, commercial, and industrial electrical work. This code is not, as it notes, an instruction manual for untrained persons. Compliance does not necessarily mean that the end product will be efficient or suitable in all respects. The focus is on safety. It is generally acknowledged that in the use of electricity, there are potential hazards. If the installation complies with the *NEC*, it will be free of the hazards. The greatest dangers are shock and electrical fire, but there are other hazards as well. For example, a heavy piece of conduit high on a wall or attached to a ceiling could fall, injuring a person below. Exacting specifications as to supporting and securing metal raceways, including types of hardware and minimum spacing intervals, go a long way toward ensuring that the conduit won't come loose.

The *NEC* is administered, revised, and published by the National Fire Protection Association® (NFPA®). A new edition is released every three years. There is an extensive review and revision procedure, with committees that debate and vote on proposed changes. The committees, composed of expert professionals from throughout the industry, meet and vote to accept or reject each proposal. A draft is compiled, and the *NFPA* general membership votes to accept the document, whereupon it becomes the current edition of the Code.

The *NFPA* is a private organization, not a governmental body. Accordingly, the *NEC* as published has no legal standing on its own. It is offered up so that states, municipalities, and jurisdictions may pass legislation that enacts it into law. They may include revisions, add mandates, or delete portions of the document before adopting it as binding within their territory. Outside the United States, certain countries such as Mexico and Venezuela recognize the *NEC*. It is also used by many insurance companies, educational organizations, and overseas military bases where electrical wiring is regulated.

The *NEC* also delineates its own applicability, and this is where questions often arise. For example, electrical wiring and equipment that are owned by a utility that generates and distributes electricity are outside the *NEC*'s scope. However, utility wiring and electrical equipment that are not involved in the generation and distribution of electricity, such as in utility clerical offices, are covered. Another example is that electrical wiring and equipment that are underground in mines are not covered, but nonmine underground equipment such as lighting in a traffic tunnel is covered. As you can see, there is some legalistic splitting of hairs involved here. As far as residential electrical installations are concerned, though, all wiring and electrical equipment are covered by the *NEC*. At first, the language may seem a little obscure, but you will quickly get used to it.

## Get Your *NEC* Now

To create a compliant electrical installation that will pass inspection and remain free of hazards for years to come, the home crafter-electrician will need to refer to the *NEC*, and it is recommended that you immediately order a copy of it and refer to it when questions arise in the course of your work. Industrial and commercial electrical work, with higher voltages, hazardous areas, and more elaborate electrical distributions, often among multiple buildings, is somewhat more complex than residential work, and accordingly, the home crafter-electrician will not be concerned with a large portion of the work. Unfortunately, the *NEC* is not divided along those lines. Information on residential installations is intertwined and dispersed throughout different locations in the *NEC*. In this book, we will endeavor to separate out the portions most relevant to residential work and discuss them under appropriate headings. A more complete treatment of the *NEC* is contained in my previous McGraw-Hill book, *2011 National Electrical Code® Chapter-by-Chapter*. In that book, I reviewed the *NEC* as it applies to commercial and industrial as well as residential locations. That book is intended primarily for working apprentice and journeyman electricians who are looking to upgrade to master status and for more advanced professionals who want to review the material. The home crafter-electrician also will find much of interest in that volume.

## Some Variations

Licensing of electricians is not exclusively a concern of the *NEC*. For the most part, it is administered, within the United States, by the individual states. Some states, among them New York and Illinois, do not license electricians but instead cede the process over to the counties or municipalities. The licensing of electricians in most jurisdictions comes in a number of variations, organized according to both the level of expertise and the type of work. A typical ranking is

- **Apprentice.** No experience or knowledge is required. The individual may work only under the direct supervision of a more experienced electrician.
- **Journeyman.** May perform electrical installations and may supervise apprentices on a one-on-one basis. This ratio is not to be exceeded. The electrical inspector may check to see if there is at least one journeyman for every apprentice on the job site.
- **Master.** The overall job must be supervised by at least one master electrician. No particular ratio is required. The master electrician is not required to be on the job site at all times, but for all professional electrical work, there must be a master electrician somewhere in the picture who is willing to certify and sign off on the quality of the work.

Many jurisdictions require, issue, and administer licenses for related activities, such as fire alarm, low-voltage wiring, electric sign installation, high-voltage work, and so on. How is all this applicable to the home crafter-electrician?

There are some general principles that are nearly the same everywhere. Regardless of the type of work—home building, electrical, plumbing, septic systems, and so on—the homeowner is usually not required to have a professional license to work on his or her own primary residence. *Primary* is emphasized. It is not to be a vacation home or a spec house built to flip. An unlicensed builder may not perform electrical work on a subdivision that is personally owned for the purpose of circumventing the law.

This allows quite a lot of latitude for the home crafter-electrician because it makes it possible to do one's own wiring project without hiring a professional. Regardless, it is usually necessary to acquire a construction permit and call in for an electrical inspection prior to operation. Note also that many utilities have their own rules regarding hookup. Many will not energize a homeowner-built service unless it is called into them by a licensed professional. Lots of electrical work, however, does not involve alteration to the service, so this is not always an issue.

Proceed with caution. If you don't want to be liable for a big fine, be obligated to do extensive rework, or even be stuck with a building that you cannot legally occupy or sell, begin by checking the requirements in your jurisdiction. The best place to start is an Internet search engine. Type the name of your state, followed by "electricians' licensing board." The site will contain administrative rules and summaries that will explain what you can and cannot do. Invariably, there will be a telephone number you can call to have further questions answered.

Even if not required, you may want to hire a licensed electrician to oversee the job or to act as a consultant. And you'll want to touch bases with the electrical inspector in advance. While home crafter-electricians' installations may be legal, your local inspector could be inclined to turn the whole thing into an adversarial ordeal. Inspectors wield enormous power simply because they have the ability to withhold operational approval. Some inspectors are fine with homeowner installations. It depends largely on the inspector's mind-set and the sometimes unwritten policy of the department. Your consulting electrician undoubtedly has had a lot of experience in this area and knows how to cope with realities that may arise.

The better inspectors, those with integrity and technical knowhow, will judge your project strictly on its merits. They appreciate good work but are extremely wary of signing off on an installation that will make problems down the road. The worst nightmare, for an inspector, is to approve an installation that bursts into flames a few days later because of a wiring deficiency that should have been caught. Electrical inspectors think about this a lot. Some react to this situation by acquiring dour, unpleasant personalities. Others strive for quality work in their jurisdictions so that each job that is inspected becomes a valuable learning experience for inspector, electrician, and end user alike.

The home crafter-electrician should be aware of the fact that although compliance with the *NEC* is the main focus of the electrical inspector, there are other dynamics that play a role as well. A well-worn copy of the most recent edition of the *NEC*, with lots of highlighting and copious handwritten notes in the margins, open to the section on grounding and left at a thoughtful angle next to the entrance panel will set the stage for a successful inspection. Mind you, it will not help if you have neglected to put in the main bonding jumper, but if you have done a first-rate job throughout, it will send the right signal.

## Electrical Deficiencies

The electrical inspector's first impression when setting foot on your property will determine the tone for the entire visit, and the most important single factor will be neatness of the premises. Even areas that have nothing to do with the electrical installation should be orderly and free of refuse and broken or corroded building materials. As for the electrical installation, the inspector will make an overall visual assessment. By this time, for many inspectors, the final judgment will be close to finalized.

Boxes should be plumb and firmly mounted. Type NM (Romex) cable, widely used in most residential applications, should be routed neatly with Code-compliant minimum bending radii and no twists (Figure 2-1).

The inspector's initial impression weighs heavily, but it is not

### Type NM Cable

Type NM cable (Romex) is permitted in one- and two-family dwellings and attached or detached garages and storage buildings. It is not permitted for backyard-mechanic garages where vehicles are repaired, even if not for pay. It is not permitted in a storage-battery room such as used for wind and solar photovoltaic (PV) installations. It is also not permitted in poured concrete. It may be exposed or concealed. It is to be supported, usually by staples, at 4½-foot minimum intervals; within 12 inches of every outlet box, cabinet, or fitting; and within 6 inches if there is no strain relief, as in a plastic wall box. Flat cable is not to be stapled on edge. Type NM is permitted to be unsupported where fished through concealed spaces. As with all wiring methods, the minimum bending ratio must be observed. For Type NM cable, the minimum bending ratio must be not less than five times the diameter of the cable, so don't wrap it around the edge of wooden framing.

FIGURE 2-1    Romex cable is acceptable for low-rise residential applications.

totally definitive and is, in fact, subject to reversal. Undoubtedly, the inspector has seen neat-appearing installations that on further examination are found to contain serious flaws. Here are some common deficiencies that electrical inspectors invariably notice:

- Enclosures and light fixture canopies that are not tightly mounted. If they can be wiggled, they will become worse in time, and eventually, the wiring will be stressed and damaged.
- Securing hardware intervals that are not observed. Details for each type of cable and raceway are given in the respective articles in *NEC* Chapter 3.
- Inadequate workspace around electrical equipment that may have to be adjusted or maintained.
- Improper location of an entrance panel or overcurrent devices.
- Directory at an entrance panel that is not complete.
- GFCIs and AFCIs that are not provided where required.
- Separate small appliance and laundry circuits that are not provided.
- Outdoor receptacles at front and rear entries that are not provided.
- Installation of a satellite dish, telephone, cable TV (CATV), or other low-voltage equipment without proper grounding and bonding.

- Improper connection of the equipment-grounding conductor to the system neutral. This should be done only once, and it should be at the service enclosure by means of the main bonding jumper.
- Improper grounding of frames of electric ranges and clothes dryers. Prior to 1996, ranges were permitted to be grounded by connection to the neutral conductor. Currently, a standard equipment-grounding conductor (fourth wire) must be run to serve this purpose.
- Failure to properly connect the ground wire to electrical devices such as switches and receptacles.
- Failure to provide a spare neutral conductor for future use in switch-loop boxes.
- Failure to install a second ground rod where required. The distance between ground rods should be at least 6 feet so that the resistance areas do not overlap.
- Failure to bond the equipment ground to a water pipe, as shown in Figure 2-2.

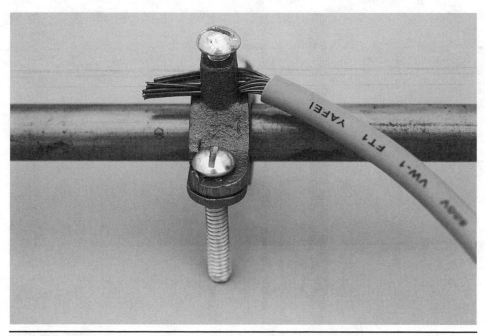

**Figure 2-2**  To make the required bonding connection to a water pipe, a proper clamp must be used.

# CHAPTER 3

# Basic Electronics
# You Should Know

S ome electricians are able to ply their trade without an extensive knowledge of basic electronic theory. After years of experience, usually working with others in the trade, they have become very good with tools, have become rather skilled in related trades such as plumbing and carpentry, have a steady hand and good eye for a neat installation, and so on. With the aid of a hand calculator, they can size out circuit conductors and raceways, and they can page through the *National Electrical Code* (*NEC*) to find specs for all kinds of methods and materials.

Many of these workers turn out safe, reliable products, often doing advanced industrial work on a regular basis. But the truth is that they are hampered by a lack of theoretical knowledge. And the same is true for the home crafter-electrician. If you acquire a knowledge of fundamental electronic dynamics and relationships right at the start, you will be on a much better track to building it right the first time, and you will be better able to troubleshoot and diagnose those "tough dog" equipment failures.

## An Invisible Domain

What is an electron? It is an elementary particle. Unlike a proton or neutron, as far as we know, it cannot be further subdivided. Protons and neutrons are made up of quarks, but electrons do not appear to be made up of anything smaller or more basic.

Electrons are very small—much smaller than protons and neutrons. Protons and neutrons bind together to form the nucleus of an atom, whereas electrons travel in orbits around this nucleus. The orbits of the planets in our solar system, including Earth, all lie in the same plane, resembling a flat disk. Electrons travel around a nucleus in ever-changing planes that are inclined with respect to each other, so these orbiting electrons are best visualized as inhabiting concentric shells that are discrete distances from the nucleus.

Atoms comprise different elements depending on how many electrons there are in orbit around the nucleus. If there is only one electron, the element is hydrogen. If there are two electrons, it is helium.

## The Meaning of Valence

The elements are arranged by atomic number (number of electrons, usually equal to the number of protons or the number of neutrons) in the *periodic chart of the elements*, and this chart tells us a lot about their properties.

The outer shell is different from the other shells. It may contain anywhere between one and eight electrons, whatever number is necessary in addition to the other shells to make up the total atomic number. The outer shell is called the *valence shell*, and it is responsible for many of the properties of an element. This is so because the electrons in the valence shell are less tightly bound to the nucleus in comparison with the electrons in the other more inner shells.

Remember that all atoms are in constant motion unless the temperature is absolute zero degrees on the Kelvin scale (0 K), in which case there is no motion of the atoms with respect to one another. If there is any heat, the atoms are in constant motion, often colliding billiard-ball-style with one another and, if it is a gas or liquid, with the inside walls of their container.

When two atoms of the same or different elements approach close enough to each other, they interact in various ways depending on the number of electrons in their valence shells. A large number of combinations and interactions is possible under different pressures and at different temperatures, and this makes up the vast field of chemistry. In the study of electronics, we are concerned primarily with the interactions that make for the behavior of a few semiconductors, such as silicon and germanium, and the activities of electrons when they become free of their valence shells, roaming the vast spaces between atoms and flowing in complex and yet orderly ways within conductors.

For the home crafter-electrician, it is not necessary to travel too far afield, yet it is worthwhile to have some understanding of these wonderful patterns of matter and energy in our universe. In the discussion on basic electronics that follows, we'll try to strike a balance.

First, a few definitions:

- An *ampere* (*amp*) is the measure of the amount of electric current that is flowing through a circuit at any given moment. Specifically, it is the amount of current flowing through a conductor when $6.25 \times 10^{18}$ (this is scientific notation for 6.25 followed by 18 zeroes!) electrons pass any given point per second. The water analogy is often useful in understanding electrical circuits: current in amps is similar to gallons per minute of water flowing through a pipe.
- A *volt* is a measure of the electrical pressure on the flow of electrons. Although amperage is an absolute measure based on a certain number of electrons and a unit of time, voltage is a derivative concept. The definition of a volt depends on the definition of an amp. One volt is defined as the amount of electrical pressure required to force a current of one amp through a resistance of one ohm. Voltage resembles water pressure in a pipe as measured in pounds per square inch.
- It follows that an *ohm* can be defined as the amount of resistance (opposition to the flow of current) there is in a load when one amp of current flows through it at one volt of electrical pressure. The unit of resistance is also derived from amperage, which itself is not derived but absolute.

## Reactance and Impedance

*Capacitive reactance* is a measure of opposition to the flow of current in a circuit that has capacitance, and it varies with the frequency of the voltage and current in the circuit and the capacitance of the load. *Inductive reactance* is a measure of opposition to the flow of current in a circuit that has inductance, and it varies with the frequency of the voltage and current in the circuit and the inductance of the load.

Capacitive reactance and inductive reactance are both measured in ohms. They conform to *Ohm's law* and behave like resistance in a circuit, although their properties are more complex. The values of these two types of reactance are frequency dependent, unlike resistance, which for the most part remains unchanged regardless of frequency.

*Impedance*, a very useful concept, is also measured in ohms. It is made up of resistance, capacitive reactance, and inductive reactance, although these are not simply added in a linear fashion. Capacitive and inductive reactance, insofar as they comprise impedance, cancel out each other, and the result combines with resistance to make up impedance.

The precise details for calculating capacitive and inductive reactance are not essential to the home crafter-electrician. The electronic technician, however, uses these operations on a daily basis.

We will need to become familiar with Ohm's law and its derivative equations, however, and this is discussed later. Before we get into that, let's look at one more definition: a *watt* is a measure of electrical power that is produced by a source such as a battery or generator or consumed by a load. One watt (also a derivative concept) is the amount of electrical power manifest by one amp of current that is driven by an electrical pressure of one volt.

Strictly speaking, electrical power as measured in watts [frequently kilowatts (kW), thousands of watts] does not flow in a circuit like amps. Instead, it is transferred from one location in a circuit to another—from the source to the load. The load, if it is a motor, converts most of the electrical power into circular motion. Because a motor is not 100 percent efficient, a fraction of the power is lost, dissipated into the surrounding space in the form of heat. Power never goes away—it just takes different forms. This is a useful concept in troubleshooting electrical circuits and equipment. Always look for the power flow. Power is best depicted by means of a one-line block diagram, with arrowheads to indicate direction.

Current flows through a circuit. Unless there are parallel divergent paths, the amount of current in the circuit is everywhere the same, including inside the source and inside the load. This is a consequence of *Kirchoff's current law*, and it is a very useful troubleshooting concept. Current is measured with an ammeter, and to take this measurement, you have to break open a conductor and insert the ammeter in series with the source and load. For this reason, an ordinary multimeter can measure only current that is in the milliamp range. The entire amount of current has to pass through the instrument. Any greater amount would quickly burn up the meter, including the probes.

Current can be measured without cutting open the circuit. Using a clamp-on ammeter (Amprobe), as shown in Figure 3-1, you can read up to 200 amps because you are measuring the current indirectly, reading the strength of the magnetic field that surrounds the conductor.

Voltage does not flow through a circuit. It is a potential difference between two points in a circuit. For there to be a reading above 0 volts, there must be an impedance between the two points. In a live circuit, you can test a switch by placing voltmeter probes on the two terminals. If the switch is on, there will be no voltage. Switch it off, and full-circuit voltage is displayed. With the circuit pow-

**FIGURE 3-1**    A clamp-on ammeter reads the current drawn by a portable power tool.

ered down, you can test the switch using an ohmmeter. Place the two probes on the terminals. When the switch is off, you should read very high ohms depending on the range setting of the meter. With the switch on, you should read 0 ohms. Wiggle the handle sideways, and if the reading fluctuates, the switch is bad. If there may be parallel resistance, preventing a true reading, you have to remove the switch from the circuit. Disconnecting just one wire will do it. The voltmeter test is the more professional way to test a switch. Many devices, such as fuses, relays, transistor output circuits, thermostats, and so on, are actually switches, and may be tested as described.

Only a minute amount of current is drawn by a voltmeter. It is safe to read 240 volts using a standard electrician's voltmeter, provided that it is set to the correct range. Many multimeters can safely read 600 volts. Above that, testing becomes tricky because other issues arise, such as whether the probes can be handled safely.

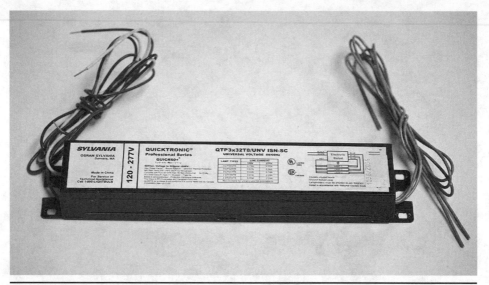

FIGURE 3-2    A fluorescent ballast.

The home crafter-electrician does not need to read more than 240 volts. Fluorescent bulbs operate at higher voltages, required for ionization, but these are not usually measured. When a fluorescent fixture fails to light, you normally replace first the bulb(s) and then the ballast, as shown in Figure 3-2, and that's all there is to it.

We have discussed amps, volts, and ohms. Now you are 50 percent of the way there. You know the meanings of the three basic electrical circuit parameters. Now we'll talk about the fundamental mathematical relations among these metrics, and when we are done, you will possess the theoretical background information needed to understand, diagnose, and repair ordinary house wiring circuits. The rest of this book will be easy to follow.

## Ohm's Law

Amps, volts, and ohms relate to one another in accordance with this formula:

$E = I \times R$

where $E$ = volts (electromotive force)
    $I$ = amps (intensity)
    $R$ = ohms (resistance)

This is the most basic formula, and the others derive from it. Actually, you don't often use this variant because you rarely need to solve for volts. The voltage is usually known because you know the system voltage supplied by the utility. For almost all residential electrical work, it is 120 or 240 volts depending on which of these you are connected to at the entrance panel. Nevertheless, it is best to know this version because it is easy to remember, and the others can be easily derived from it. You will only need to memorize this one formula, and the other two will be readily available.

When there is an equation such as the one just cited, you can perform the same operation on both sides, and the equation will remain valid. You can also switch the two expressions on either side of the equal sign, and this is usually done for ease in reading so that the single unknown is by itself on the left.

By dividing both sides by $I$, we get

$E/I = R$

which is the same as

$R = E/I$

Using this formula, we can find the resistance in ohms when we know the volts and amps in a circuit.

Similarly, by dividing both sides by $R$, we get

$E/R = I$

which is the same as

$I = E/R$

Using this formula, it is easy to find the current in amps when we know the voltage and resistance in ohms in a circuit. This variant is frequently used because it is always necessary, in an electrical circuit, to know the current that will flow through it in order to size out the conductors and any devices such as switches and relays so that they can carry the load safely.

There is one other formula that, while not actually part of Ohm's law, is closely associated with it and is frequently used by electricians on any size job. That is,

$P = E \times I$

where $P$ = watts (power). Deriving from this formula, we get

$E = P/I$

which is rarely used, and

$I = P/E$

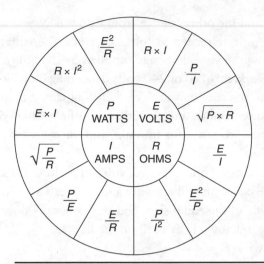

**Figure 3-3**    Ohm's law wheel displays Ohm's law and the power equation in all forms for instant access.

This variant is used very frequently, such as when the voltage and power of a load such as a hot water heater are known and it is desired to find the current in order to size out the supply circuit including conductors and overcurrent device.

Ohm's law and the power equation interact to yield some other equations, and there is a wonderful graphic that brings all this together. It is called *Ohm's law wheel* and is shown in Figure 3-3.

If you have assimilated the information introduced so far in this chapter, you have all the theoretical knowledge you need to design, install, troubleshoot, diagnose, and repair ordinary house wiring. There are a few other calculations you will run into from time to time, but for the most part, that knowhow will be derived and follow from the preceding.

Of course, we have to stress that this theoretical background, although important in itself, is only a start. There is a lot to learn about the various wiring devices and how they go together; use of test equipment (especially the multimeter); planning, laying out, installing, and repairing concealed wiring in a home; and some of the extras that you may encounter in home wiring, including data, telephone, and other low-voltage work. Above and beyond this, there is home automation, backup power (including transfer switches), and solar and power cogeneration with synchronous inverter hookup.

## Getting Started

Even when you draw a circle so as to exclude everything that is not residential, this is still a big subject. But it won't be too difficult if you go one step at a time. Begin with some simpler jobs, such as wiring receptacles and switches and doing some home runs to the entrance panel. At this stage, it is a good idea to work with a professional. (It may be feasible to pay $15 and get an apprentice card. Then you can learn on the job.) Before you know it, you'll be wiring the box, building the service, installing light fixtures, and hooking up three-way switches. In this book, we'll be looking at these and similar projects. We won't waste time on feckless discussions about whether the ground prong goes on top or bottom, and we'll try to refrain from presenting too much detail about harmonic distortion, magnetic resonance, and the like. The goal is to stick to residential wiring and cover as much detail as possible in a single book.

## Concealed versus Exposed Wiring

Residential wiring is simpler than commercial or industrial work because it is smaller in scope, there are fewer voltage and current levels with less arc-flash hazard to worry about, and the connected electrical equipment is less complex. In one respect, however, residential work can be more difficult because a better finish appearance is usually necessary. Because most home wiring is concealed, there are accessibility issues that do not arise on the factory floor. Once the walls are filled with insulation and the wiring is covered by wall and ceiling finish, it becomes more difficult to do alterations and repairs because the cabling cannot be easily removed and replaced. In a commercial or industrial environment, even if raceways are behind wall and ceiling material, it is a simple task to install a new cable run using the old wire as a pull rope.

The home crafter-electrician must become adept at concealing wire for the sake of appearance, and where alterations or repairs are being made, this becomes high art. The *NEC* permits the familiar Type NM cable (Romex) to be stapled to the wall finish, but this is acceptable only in a rustic cabin or unfinished garage. In finished offices and stores, the problem is solved by running wiring above suspended ceilings. Panels can be easily popped out to access the cable so that it can be altered and repaired as needed, and new wiring can be added. Suspended ceilings often are not considered acceptable from an aesthetic point of view in residential living rooms and bedrooms, so we are back to the problems inherent in concealed wiring. One solution, for a retrofit, is to use Wiremold. This is a metal raceway that has a nice finish and is suitable for use on finished surfaces. It comes with a complete line of fittings, enclosures, and devices and very good installation instructions. Many sizes, shapes, and colors are available. It is installed like any raceway, and then conductors are pulled through it. However, it adds to the expense of the job, so it is better to conceal the wiring in the first place where possible.

## Residential Work

Now we'll take a tour through a typical residential electrical installation beginning upstream. We won't say too much right now about the service because Chapter 4 is devoted to that topic. Suffice it to mention that the service consists of that portion of the premises wiring starting at the utility point of connection and ending at the input terminals of the main overcurrent device, which may be in the entrance panel or in a separate main disconnect enclosure, either inside or outside the building.

The utility often requires that the service be built or at least certified by a licensed electrician, but in any case, the home crafter-electrician should be familiar with this part of the electrical structure because it contains the grounding means and constitutes the jumping-off place for the entire premises wiring. The usual procedure for wiring a house is to begin by mounting the enclosures (with knockouts removed) in place. Then, if the house is wood framing, drill the holes in the studs and framing members using the correct size drill bit. Pull the wires through the holes, staple them to the framing at *NEC*-specified intervals, and insert the cable ends through the connectors, leaving sufficient free wire at both ends. At the wall boxes, the *NEC* specifies that 6 inches of free conductor beyond the inner rim of the enclosure be left for making connections. Some workers cut the ends shorter in the belief that it will reduce box fill, but this is a mistake because it is more difficult to make good terminations.

## Wiring the Box

At the entrance panel, you need enough length of the black conductor to reach the breaker. The white conductor must be long enough to reach its breaker for a 240-volt circuit or the ground bar for a 120-volt circuit. The bare or green equipment-grounding conductor has to be long enough to reach the grounding terminal. In all cases, leave your whips long enough so that they can be pushed back into the corners of the box. Make all bends right angles, as opposed to taking shortcuts through the available space. In this way, the first few circuits won't overfill the box, making it difficult to add others.

When it comes time to make terminations, if you find that one of your wires is too short, it is acceptable to make a splice using wire nuts inside an entrance panel. It is better, however, to leave enough free conductor in the first place.

In new construction, it is best to put in the service and heat up the entrance panel at the outset. In this way, there is power to work with, and the temporary service can be removed.

The entrance panel can be in the basement or upstairs, conceivably on the second floor. The *NEC* prohibits the installation of overcurrent devices (hence entrance panels) in bathrooms, clothes closets, and on stairways, but outside of that, it is your choice. We'll have much more to say about the location of entrance panels in Chapter 4.

It is best to wire the service entrance conductors into the entrance panel before terminating the branch circuits. In this way, the branch-circuit conductors are not blocked, and they can be shifted around later if the need arises.

The usual location for the entrance panel is in the basement. Then the branch circuits and feeders can be run along joists or sills. They remain accessible for troubleshooting purposes or if changes must be made. Cabling can be run anywhere in the basement and stubbed up through the floor. A centrally located chase inside an interior wall can be built to bring cable runs to the second floor.

If holes are drilled in load-bearing framing members, they should be as small as possible to permit easy installation of cable. A ⅝-inch hole is suitable for 12 American Wire Gauge (AWG) Romex. Large holes should be avoided because they weaken the framing members. A hole drilled near the middle of a long span will weaken it more than if the hole is drilled closer to where the span is supported. A long floor joist will tend to sag near the middle. The top edge of the timber is in compression, and the bottom edge is in tension, that is, trying to stretch. The center is neither in compression nor in tension. Therefore, any holes should be drilled near the center so that there is less tendency to weaken the framing member. This has the added advantage of providing more isolation in regard to nail penetration.

If the entrance panel is to be mounted on a concrete wall, the usual procedure is to make a ¾-inch exterior-grade plywood backing panel that should be about 10 inches wider than the box all around so that cables can be stapled in place. All bends should be 90 degrees for a neat appearance, with the turns gentle enough to comply with the minimum bending radius for the type of cable, as specified in

*NEC* Chapter 3. This work should be precise and neat in order to impress family members and visitors and to facilitate any future wire tracing. It is customary to paint the backing board a low-gloss black.

The *NEC* has some requirements regarding installation of the entrance panel. If it is mounted directly on the concrete wall, it must have standoffs so that there is at least ¼ inch of air space between the box and the wall. Wooden dowels are not to be used in drilled holes in the masonry.

Most entrance panels use circuit breakers, as shown in Figure 3-4, for overcurrent devices. Fuses are permissible, but they are rarely used in this application because in the course of interrupting the circuit when subject to overload or short circuit, the element burns up, and they must be discarded. Breakers can be reset, and moreover, a breaker box is easier to wire and maintain.

A single-phase breaker box, as shown in Figure 3-5, consists of a metal enclosure housing two current-carrying metal bus bars that are insulated from each other and from the metal enclosure. They extend down most of the length of the

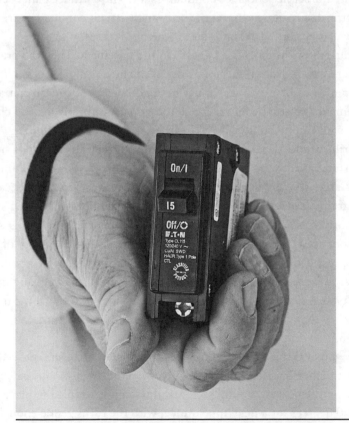

**FIGURE 3-4**   A 15-amp single-pole breaker used in many 120-volt branch circuits throughout a residence.

FIGURE 3-5   A 200-amp entrance panel with double-pole main breaker and main bonding jumper.

box, leaving space below for wiring to cross over. At the top, the main breaker fastens to the bus bars. The main breaker is a double-pole breaker with large input lugs for the service-entrance conductors. In the usual single-phase residential system, the voltage between these two lugs measures 240 volts, as does the voltage between the two bus bars. Between either of these and the neutral bar, panel enclosure, or anywhere along the grounding system, the voltage measures 120 volts. This is what is known as a *three-wire, two-voltage system*. It follows from the fact that the two hot legs originate at the endpoints of the secondary winding inside the utility pole– or pad-mounted transformer, whereas the neutral conductor originates from a center tap of this same winding. When you put your hand on a grounded water faucet in the home, you are putting your hand on the center tap of the secondary winding within the utility transformer and on similar metal at the substation and even at the generating plant. Harmful voltages are not felt because of the low-impedance grounding at numerous points throughout the system.

While you are installing the entrance panel, there are a couple points to always keep in mind:

- **In a service entrance panel, the neutral bar must be bonded to the metal enclosure.** The connection is to be made by means of the main bonding jumper, which is attached at the time the box is installed. The reason that it is not part of the box as it comes from the factory is that it is not always required. In fact, it is prohibited when the box is used as a load center downstream from the main disconnect. As stated earlier, such bonding would violate the injunction against multiple bonding of the neutral and equipment-grounding conductors. This connection must be made only once, inside the service entrance panel, and never anywhere else. The main bonding jumper usually takes the form of a threaded screw attached to a card included with the entrance panel. The card reads, "Attach this main bonding jumper in the entrance panel when required." In the neutral bar, there is a threaded hole for the main bonding jumper. When the box is used as a service-entrance panel, screw the main bonding jumper tightly in place so that it cuts through the paint and digs into the metal of the enclosure, making an electrical connection. If you neglect this simple screw, your enclosure will not be bonded, and if a live wire were to chafe anywhere inside, the box would become energized, creating a hazard. Here's another very important item that is often neglected by novices: any metal water piping in or on the building must be bonded to the grounding system. To do this for a 100-amp service, use 6 AWG copper wire, solid or stranded, bare or with green insulation. For a 200-amp service, 2 AWG copper is required. Insert one end into the oversize hole in the neutral bar. Run it out through the miniature punch-out at the bottom of the enclosure, and connect it to the nearest metal water pipe using a pipe-grounding clamp made for the purpose. Using this same 6 or 2 AWG wire, make a jumper around the water meter, if there is one, so that if the meter is removed, ground continuity will be preserved. Also jump around any non-metallic housings, such as associated with water filters, and jump around any short runs of plastic piping that may separate metal segments.
- **Fill out the directory, usually attached to the inside of the cover that opens to access the breakers.** It is a Code violation to neglect the directory. The printing should be neat and legible. If you use ink, it will be difficult to erase when alterations are made in the future, as is usually the case.

Branch circuits and feeders originate in the panel. A branch circuit runs from the final overcurrent device to the load. A feeder runs from one overcurrent device to another, as in a load center. In other words, a feeder has overcurrent devices at both ends. A mistake novices make is that they place unneeded load centers throughout the occupancy, with local branch circuits emanating spider-web-fashion from each one. This is an expensive variant with no upside unless it is needed in an unusually large building to mitigate voltage drop. When a circuit trips out, it makes for more difficulty in finding the overcurrent device.

Branch circuits are individually wired into separate breakers in the entrance panel. Except for the main, the breakers are purchased separately. Square D, which makes high-quality products, and some other makers require unique breakers that are not compatible with other brands. Many makes are compatible with ITE breakers, which means that the breakers fit the mounts and clip into the bus bar correctly. However, the metal alloys may differ so that over a period of time a corrosion could be a problem. Consult the Underwriters Laboratories (UL) listing, the manufacturer, and the electrical distributor to resolve this problem.

Be sure to include an equipment-grounding conductor. Terminate it at the grounding-terminal strip in a load center or the neutral bar in a service-entrance panel. Put the breaker in place, and route the ungrounded conductor(s) to it. Cut the conductor(s) to length, and strip off just enough insulation so that there is bare wire inside the lug but no copper showing outside. Pull the breaker out, connect the conductor(s) with sufficient torque, and then replace the breaker, making sure that it is seated correctly. Each branch circuit should be completed before tying it into the box to avoid working on live wires.

Grounding conductors and grounded conductors are wired to the neutral bar or the grounding bar. More than one grounding conductor may go into a single lug, but the grounded conductors may not be doubled up. This is so because at some time in the future one of the grounded conductors may have to be removed, at which time the circuit sharing the neutral termination would lose ground continuity, destabilizing the voltage with respect to ground.

The exact height of receptacles is not specified by the *NEC*. A good height is 10 inches from the subfloor to the bottom of the wall box. Throughout any house, the receptacles all should be the same. Switch heights should be such that the switches can be operated with the forearm level. A good height is 46 inches to the bottom of the box. Thermostats, which take a standard wall box, should be eye level, 60 inches to the bottom of the box. You can make gauges to aid in setting wall boxes to the appropriate heights.

Wall boxes should extend beyond the inside of the framing so that they will be flush with the anticipated finish wall material. Some wall boxes have a mark or ridge to aid in positioning them when ½-inch sheetrock is to be used. Be sure that the wall box does not extend too far out, or the wall plate will not seat on the finish wall, leaving an unsightly gap.

There are several mounting styles for wall boxes. Choose one that works for you. A common type mounts using two 16-penny nails. Boxes are available in metal or plastic. Plastic is a little cheaper, and in a big subdivision or in the life of an electrical contracting firm, the savings would be substantial. On a single small project, the cost difference would not be decisive.

Plastic boxes do not require Romex connectors as long as the cable is stapled within 6 inches of the box. Metal boxes dissipate heat better in case of an arc fault, and they do not contribute to the overall fire load. If home automation is contemplated for the project, plastic wall boxes are better because they make for less radiofrequency (RF) shielding where wireless connectivity is to be used. Both plastic and metal boxes are acceptable if they are UL listed.

When you put in a ground-fault circuit interrupter (GFCI) or if there are multiple runs terminated at a single box or a run going out to a switch loop with wire nuts, it may be very difficult to put the device into the box after it is wired. The solution for this problem is to buy and use deep wall boxes. Some big electrical contractors, to save a few cents on every job, use a shallow wall box for the last receptacle on each branch circuit and for switch loops.

In new work, the usual procedure is to bring the cable into the wall box, tighten the connector (not too tight; just snug so that the cable can't slide), and then form the wire into a coil and put it into the box so that it won't interfere with the sheet-rocking process. This completes the job of roughing in the wiring. Later, after the walls are taped and painted, it is time to do the finish electrical work. This consists of pulling the cable ends out of the wall box, slitting the jacket back to about ½ inch from the connector, removing the paper strip, stripping the insulation off the ends of the current-carrying conductors, and wiring the devices.

## Wiring the Devices

It is possible to strip the insulation from a conductor using a utility knife with a new sharp blade. Whittle off the insulation as you would sharpen a pencil with a knife. Under no circumstances can the copper be nicked. If that happens, when the circuit is heavily loaded, there will be a hot spot in the wire right next to the termination—and that is a fire waiting to happen! If you nick a wire, cut it back, restrip, and if necessary, connect a short jumper.

Rather than using a knife, a wire stripper does a better job. The automotive type works, but it is bulky and difficult to get into tight places. The professional electrician's stripping tool is perfect for this job. When stripping the end of a conductor, look at the termination and carefully judge how much insulation to remove. This is critical. The idea is to remove enough insulation so that none of it will get caught under the screw, which would compromise the electrical connection. On the other hand, if you remove too much insulation, there will be exposed copper, which could arc to ground or be a shock hazard.

Some devices have a back-wiring option. The stripped end is inserted into a hole in the back of the device, where it is held in place by spring tension. This type of termination does not have as great ampacity as a good screw termination that is torqued properly. There is rarely a good reason for using the back-wiring option.

It goes without saying that receptacles have to be polarized properly. The grounded (white) conductor is connected to the screw that has a silver finish, and the black conductor is connected to the screw that has the brass finish. Devices that have holes for terminations have the word *White* or the letter *W* for the grounded conductor terminal. Devices with separate inputs and outputs, such as GFCIs, are marked with the words *Line* and *Load*.

When wiring a residential occupancy, whether it is an entire new building, an addition, or an outbuilding such as an attached garage, the object should be to create a Code-compliant product, and one of the important tasks is to have the

> ### Receptacles versus Outlets
>
> Often customers in a hardware store ask for a dozen outlets. The proper term is *receptacles*. An *outlet* is any device or equipment that is attached to a branch circuit so that it may be powered up. A hard-wired light fixture is an outlet. So is a receptacle. A *receptacle* is a device, often duplex, into which an attachment plug may be inserted. A receptacle on a store shelf is a device. Wired into a branch circuit in the home, it becomes an outlet.

correct switch and receptacle placement. Code requirements are exacting. For example, in habitable rooms such as living rooms and bedrooms, there is a prescribed maximum spacing between receptacles. But how do you handle doorways, large archways, glass sliders, and the like? You have to know where in the *NEC* to find these mandates. They are in Chapter 2, Article 210, "Branch Circuits," not Chapter 4, Article 406, "Receptacles, Cord Connectors, and Attachment Plugs." These requirements are maximum intervals, so you are free to install additional receptacles.

Nonelectricians sometimes ask what maximum number of receptacles is permitted on a single branch circuit. This is not the way it works. Seven is sometimes mentioned as a rule of thumb, but adding more receptacles does not increase the load, and it is permitted without limit. Additional receptacles provide more locations for cord and plug connections. The primary hazard is not overloading the circuit because the overcurrent device will take care of that. The primary hazard is overuse of extension cords, so, within reason, the more receptacles the merrier.

## Another *NEC* Violation

While we are on the subject, flexible cords are not to be used as a substitute for permanent wiring. Cords should not be stapled along a wall or run through holes in walls, floors, or ceilings. Wherever a cord is plugged into a receptacle, it must be in sight, not in walls, cavities, above suspended ceilings, or anywhere that is concealed from view.

In nonresidential occupancies, such as stores and factories, there is no mandate on receptacle spacing or minimum number. Receptacles are provided for the anticipated need. In a residence, however, there are definite specifications.

In kitchens, there must be two 20-amp small-appliance circuits. They are usually placed along the counter space. Those within 6 feet of the rim of a sink must be GFCI protected. Even though the circuits are rated 20 amps, the receptacles may be 15 amps. This is an exception to the general idea that the overcurrent device is the weakest link in the chain. Permitted receptacle ratings are listed in Table 3-1.

Contrary to common belief, a refrigerator does not need a 20-amp circuit, nor is a GFCI required or even desirable. GFCI protection, generally speaking, is incompatible with refrigeration because loss of power, unnoticed, will equate to food spoilage. Moreover, refrigeration equipment has a tendency to experience nuisance tripping because windings in the hermetically sealed motor-

**TABLE 3-1**   Maximum Receptacle Ratings for Various Size Circuits

| Circuit Rating (amps) | Receptacle Rating (amps) |
|---|---|
| 15 | Not over 15 |
| 20 | 15 or 20 |
| 30 | 30 |
| 40 | 40 or 50 |
| 50 | 50 |

compressor, immersed in fluid, will ground out when the insulating coating bonded to the wire begins to deteriorate. This happens because any slight water contamination in the refrigerant reacts to form an acid that etches down to the copper. Old but still functional refrigerators exhibit enough leakage current to ground to trip out the GFCI. Like all non-double-insulated electrical equipment, refrigerators should have an intact equipment ground to prevent metal parts from remaining energized.

Along the countertop, the receptacles are to be spaced so that no point is more than 24 inches from a receptacle. To maintain this maximum distance, it is necessary to place receptacles every 48 inches, with a receptacle no more than 24 inches from each end or from a sink.

In habitable rooms such as living rooms, dining rooms, bedrooms, and the like, there is a similar geometry, except that the maximum distance of any point from a receptacle measured along the floor line is 6 feet, so the normal interval is 12 feet, except where runs end at breaks such as archways and glass sliders, from which the maximum distance is 6 feet. Complete details are given in *NEC* Article 210, "Branch Circuits," which should be reviewed, paying close attention to the wording, prior to beginning this phase of the work.

## Switches

Switches are a vital part of any electrical installation, and every home has many of them. They should be provided as the *NEC* mandates, wired correctly and located for maximum convenience for the end user.

Every habitable room should have a ceiling light. It must be controlled by a switch on the inside wall on the knob side, not the hinge side, of the door. Roughing in the wiring, the electrician should find out which way the door will swing. If there are to be other switches at this same location, they should be grouped in a single two- or three-gang wall box. In this type of configuration, the switch nearest the door should control the light fixture for the convenience of the end user entering the darkened room. The other switches, if there are more than one, should be arranged in some kind of logical order. The *NEC* permits, where for any reason it is desired not to have a ceiling light fixture, that this switch may control instead a dedicated receptacle that will supply power to a lamp.

Switching may be in either of two configurations. One is the in-line switch, and the other is the switch loop. The one you choose depends on the layout of the room with regard to the location of the power source, switch, and load. You need to decide which alternative is more economical of wire, and that one usually will require less installation labor. If the switch is between the power source and the load, the in-line configuration is better. If the load is between the power source and the switch, a switch-loop configuration is used. In both instances, we are talking about locations along the wire run, not necessarily spatial relations.

To wire an in-line switch, bring cable from the source to the switch. This is called *live power*. Then run cable from the switch to the load. This is called *switched power*.

A load could be switched by breaking the grounded conductor (neutral, white), but this would be a Code violation and very dangerous. The load would be turned off, but the ungrounded (hot, black) conductor and internal circuitry would be live. This would create a shock hazard for a maintenance worker who would assume that the equipment is powered down. An in-line switch always should be placed in the grounded conductor. And, of course, the equipment ground is never to be switched.

If the load is 240 volts, it is powered by the two hot legs from the single-phase supply. In this instance, it is necessary to break both ungrounded conductors simultaneously, again without affecting the neutral, if there is one. (Some 240-volt loads require neutrals; others do not. Single-phase motors, baseboard heat, and hot water heaters do not require neutrals. Most electric ranges and similar appliances require neutrals because they contain 120-volt circuits such as lights and/or clocks. In no event is the neutral to be switched.)

The other configuration is the switch loop. Here the live power is brought directly to the load, such as a light fixture, from the entrance panel or load center. The grounded connector is not connected to the switch. Instead, it goes to the load. At the switch, the live power ungrounded conductor is connected to the switch. It is customary to connect this wire to the bottom terminal for consistency and to facilitate troubleshooting and repair, but it will work the same either way.

The *NEC* requires that a neutral be present in every switch enclosure. It is not necessary for a simple switch to operate. In the case of an in-line switch, it is already there, but for a switch loop, an extra neutral must be provided for future use. To comply with this rule, 14-3 AWG Romex is generally used. In this way, there is an extra neutral (white), the hot supply to the switch (black), and a return hot conductor (red) back to the load. This wire is connected to the hot terminal of the load. The spare neutral is tapped from the neutral line within the load enclosure. The purpose of the extra neutral run to a switch that is on a switch loop is so that if sometime in the future it is decided to upgrade to home automation or energy-saving electronics that require power, it is available. The in-line configuration is more economical because you don't have to bother with the three-wire cable, so it is often the better choice even when the layout of the room would seem to point to a switch loop.

## Wiring Three- and Four-Way Switches

If it happens that a room has two entries, say, at opposite ends, special arrangements are needed. If there were two separate standard single-pole, single-throw switches, one at each end, this would not be satisfactory. If the two switches were in series (like a digital AND gate), both would have to be on to light up the room. If the two switches were in parallel (like a digital OR gate), both would have to be off to turn off the light(s). Either way, there would be instances where the user would have to cross the room in darkness to control the room lighting.

The solution to this dilemma is the three-way switch circuit. This ingenious arrangement allows the user to control the load from either of two locations. The addition of four-way switches permits control from any number of additional locations. Other applications for three-way switch pairs include stairways, outbuildings so that lights can be controlled from inside either building, attached garages, outdoor lighting including porch lights so that it can be controlled from the house or an outbuilding, and so on.

Many individuals have problems wiring these switches together with the source and load and having the final product work correctly. They have to call in a professional to straighten out the terminations and/or cable runs. Even some experienced electricians, if they haven't done three-way switches in awhile, have to learn them all over. The whole thing becomes simple and easy to remember if you keep a few basic principles in mind.

There are two basic situations with subdivisions. One is the in-line configuration, and the other is the switch-loop configuration. The subdivisions involve whether power from the entrance panel or load center is initially furnished to either the first or the second three-way switch or to the load.

Three-way switches are specialized devices. The handle has no marked on or off position because this varies depending on the state of the other three-way switch. It has three terminals, all colored brass because there is never a neutral (white) connected to it. On one end of the body is a single terminal, marked "common." On the other end of the body are two terminals that are not marked. Electricians call the conductors that are connected to them *travelers*.

If you have a three-way switch on hand, set your multimeter to the ohms function, and ring it out. You will see that regardless of the position of the handle, there is never continuity between the traveler terminals. Between the common and one of the traveler terminals, there is continuity when the handle is thrown one way and no continuity when the handle is thrown the other way. Between the common and the other terminal, there is continuity only when the handle is thrown the opposite way. In other words, by throwing the switch alternate ways, you can connect the common to either of the two traveler terminals one at a time but never to both simultaneously.

The two three-way switches are mounted in the two locations where switching of the load is desired. From the power source, bring a run of 12-2 AWG Romex into the first three-way switch wall box. Do not connect the ungrounded (white) conductor to the switch. Wire-nut it through to the other three-way switch wall box, where it is similarly wire-nutted through to the load. Electrically, the white is connected only to the power source and to the load, never to a switch. The ungrounded (black) conductor from the power source is connected to the common terminal of the first three-way switch.

Between the two traveler terminals of the first three-way switch and the two traveler terminals of the second three-way switch, a different type of cable is used. By way of background, 14 AWG conductors are permitted for many household loads because the ampacity is sufficient. Other loads require 12 AWG conductors. Many electricians, to simplify inventory and just because they think it is better, use 12 AWG for all lightweight branch circuits, even where 14 AWG would suffice. An exception, however, is for switch loops, where three-wire cable is needed, load permitting. (This cable actually has four wires if you include the equipment ground, but for this nomenclature, it doesn't count.)

Between the two three-wire switches, run 14-3 AWG Romex. This cable consists of one each of white, black, red, and, as always, a bare or green equipment-grounding conductor. The white is for the grounded neutral, and the red and black are the travelers. (I call them *politicians* in order to introduce a little humor into an otherwise dry topic.)

Black and red are insulation colors indicating that the conductors are ungrounded or hot. Actually, any color can be used to denote ungrounded conductors except for white, which is reserved for a grounded conductor, and green, which is reserved for a grounding conductor. If you were running raceways including Wiremold, you could pull a green and a white, with yellow and blue, for example, for the travelers, but with Romex, you are limited to what is available, which is white, black, red, green, or bare.

The 14-3 AWG Romex is the only cable between the two three-way switches. At each end of the 14-3 AWG Romex, the black traveler is connected to one traveler terminal, and the red is connected to the other. It does not matter which is which or if they cross over.

From the second three-way switch, that is to say, the one that is farther from the power source and closer to the load, run 12-2 AWG Romex to the load. At this switch, connect the black conductor to the common terminal, and wire-nut the white conductors that just pass through the two three-way switch enclosures with no electrical connection to either switch. At the load end, which is usually a light fixture that has one black and one white lead, wire it up the conventional way, color to color (this is electricians' jargon that means white to white and black to black).

This covers the most basic circuit for a pair of three-way switches when they are in the in-line configuration; that is, both switches are located (electrically) between the power source and the load. The best way to visualize the pair of three-way switches is as a single *black box*, the travelers and interiors of the two switches being inside this conceptual box, with the common terminals of the two three-way switches mounted on the outside of the box. This single unit functions as a standard single-pole, single-throw switch. There is either continuity between the two common terminals, in which case the load is powered up, or there is no such power to the load. The two three-way switches should be regarded as positioned so that the traveler terminals are facing (electrically) one another. The common terminal of the first three-way switch faces the power source, and it is the input to the pair of switches, seen as a unit. The common terminal of the second three-way switch faces the load, and it is the output.

If it is desired to have more than two locations that will be capable of controlling the load, additional three-way switches won't work. The way it is done is to insert one or more four-way switches in the travelers' line. There is no limit to how many can be used. As the name suggests, a four-way switch has four terminals, a pair that is the input at one end of the switch body and a pair that is the output at the other end. It doesn't matter which end is which, and you don't have to keep track of the travelers. They may cross over any number of times between the two three-way switches. Four-way switches are easier to wire than three-way switches. Just remember that the white wires are wire-nutted straight through from power source to load, with no connection to any of the switches. It's helpful to ring out a four-way switch using your multimeter in the ohms mode.

## Three-Way Switch Loop Configuration

Three-way switch loops accomplish the same results with somewhat different circuit wizardry. A switch loop is useful in a situation where it is more economical to run power from the source directly to the load and from there cable down to the three-way switches and any four-way switches along the way. Connect the neutral (white) of the 12-2 AWG from the source directly to the neutral terminal of the load. Do not connect the ungrounded conductor to the load. Instead, wire-nut it through the nearer three-way switch and onto the farther three-way switch, where it connects to the common terminal, becoming the input to the black box.

Here's the part that causes difficulty: if the whites in the loop are reidentified and used as return hot wires, this leaves us with no conductor to provide the neutral for future use. Therefore, it is necessary to run a 14-3 AWG wire from the load to the first three-way switch and a 14-4 AWG wire between the two three-way switches. The 14-4 AWG wire will have one white conductor and three conductors that are other than white or green. Use two of these colored conductors as travelers, one as the supply for the common that is the input and the white as the extra neutral.

What if you want to bring power from the source to the first three-way switch? Then you won't need an extra neutral there because you already have it. You will need 14-4 AWG wire to get a spare neutral, two travelers, and a return hot wire to the second three-way switch.

In view of the expense (14-4 AWG wire costs twice as much as 14-3 AWG wire), it is best to avoid switch loops. Note also that the spare coils and wire nuts increase box fill, so be sure to use deep boxes. Most of the time, in-line switching is best even if the runs are a little longer. Dimmer switches are wired using the same circuits. Three- and four-way dimmers are available.

## Sizing Feeders and Services

When it comes to sizing residential electrical work, some *NEC* navigation with table reading and number crunching is needed. We'll have a lot to say about services in Chapter 4, but for now, we'll consider the general procedure for sizing them. The point of departure is *NEC* Table 220.12, "General Lighting Loads by Occupancy," which is reproduced here:

| Type of Occupancy | Volt-Amperes per Square Foot |
|---|---|
| Armories and auditoriums | 1 |
| Banks | 3½ |
| Barber shops and beauty parlors | 3 |
| Churches | 1 |
| Clubs | 2 |

| | |
|---|---|
| Court rooms | 2 |
| Dwelling units | 3 |
| Garages, commercial (storage) | ½ |
| Hospitals | 2 |
| Hotels and motels, including apartment houses without provision for cooking by tenants | 2 |
| Industrial commercial (loft) buildings | 2 |
| Lodge rooms | 1½ |
| Office buildings | 3½ |
| Restaurants | 2 |
| Schools | 3 |
| Stores | 3 |
| Warehouses (storage) | ¼ |
| In any of the preceding occupancies except one-family dwellings and individual dwelling units of two-family and multifamily dwellings: | |
| Assembly halls and auditoriums | 1 |
| Halls, corridors, closets, stairways | ½ |
| Storage spaces | ¼ |

In the left column appear 18 types of occupancies with load per unit of area. The right column gives inch-pound numbers rather than metric. Notice that there is a substantial difference in the electrical lighting load for different occupancies. At the low end are storage warehouses at ¼ volt-amp per square foot. This is so because most of the time, for most of the building, the lights are off.

At the other end of the scale are banks and office buildings, both of which are rated at 3½ volt-amps per square foot. Dwelling units are not far behind, at three volt-amps per square foot. For dwelling units, the calculated floor area does not include open porches, garages, or unused or unfinished spaces not adaptable for future use.

For dwellings, unlike other occupancies, this general lighting load includes the receptacles, which for nondwellings after derating have to be added in separately. So what remains is to make a list of all appliances and nonreceptacle/lighting loads together with applicable rules and derating factors. This subtotal is added to the general lighting load to find the total connected load. The total is divided by the system voltage to obtain the number of amps, which determines the size of the service for a new building. For an addition, you will be able to determine whether it is necessary to upgrade to a larger service. Many existing buildings have a 100-ampere service that is filled to capacity, so a new service is always a distinct possibility.

Associated with the preceding table are supplementary provisions and explanatory material in Sections A through L, so consult the *NEC* to make sure that you have everything figured correctly. When setting up the branch circuit sizes, remember that motors and continuous loads are calculated at 125 percent of the nameplate rating, but this surcharge does not have to be added when sizing the service. Such equipment is usually manufactured in sizes with seemingly odd ratings that when multiplied by 1.25 qualify for standard-size overcurrent protection. Continuous loads are those expected to operate for over 3 hours. An example of a continuous load is electric baseboard heat. An example of a noncontinuous load is a quick-recovery electric hot water heater. For circuits supplying inductive and light-emitting diode (LED) lighting loads that have ballasts, transformers, autotransformers, or LED drivers, the calculated load is to be based on the total ampere rating of the units, not the watt ratings of the bulbs.

An important principle to keep in mind in doing these and similar calculations is that where there are noncoincident loads, only the larger of them needs to be figured in. The classic example of noncoincident loads is heating and air conditioning.

In a kitchen, recall that two 20-amp small-appliance circuits are required. They have to be added as part of the total connected load. They are figured as 1,500 volt-amps each. A laundry circuit also must be provided. It is likewise figured at 1,500 volt-amps. Both the two small-appliance kitchen loads and the laundry circuit load may be added to the general lighting load so as to be subject to the demand factor in *NEC* Table 220.42.

It is permissible to apply a demand factor of 75 percent to the nameplate rating of four or more electrical appliances fastened in place, other than electric ranges, clothes dryers, space-heating equipment, or air-conditioning equipment in a dwelling. Electric clothes dryers in a dwelling are to be either 5,000 watts or the nameplate rating, whichever is larger. There is no permitted demand factor unless there are over four clothes dryers.

## The Infamous Column C

Now we come to a more complex set of calculations, all based on Table 220.55 together with footnotes and explanatory wording. Table 220.55 is titled, "Demand Factors and Loads for Household Electric Ranges, Wall-Mounted Ovens, Counter-Mounted Cooking Units, and Other Household Cooking Appliances over 1¾-kW Rating." It is stated that Column C is to be used in all cases except as otherwise permitted in Note 3.

The table is used for most electrical work in residential kitchens unless the cooking equipment is gas fired. The body of the table gives demand factors as percentages for Columns A and B. But for Column C, it is not a percentage demand factor. Instead, it is a maximum demand in kilowatts. That is why the table looks peculiar at first glance. We are used to reading tables populated by data that are in the same units.

Columns A and B are optional. Column A may be used if the appliances are rated at less than 3½ kW. Column B may be used if the appliances are rated as 3½ through 8¾ kW. For Column C, there is no kilowatt rating minimum, and the maximum is much higher than Column B (12 kW), which is higher than you will see in residential work.

So we see that Column C may be used anywhere that Column A or B can be used. It goes higher than either column. It may seem strange that this table covers as many as 61 appliances, but this is necessary for large apartment buildings with many individual kitchens on the same service.

Before the load is totaled for the purpose of calculating the size of the service, there is one other operation that is permitted, and it has the effect of reducing the size of the service that is required. *NEC* Table 220.42, "Lighting Load Demand Factors," lists percentages that may be applied for various occupancies. We are interested in the first row, dwellings, shown in Table 3-2.

**TABLE 3-2**    Lighting Load Demand Factors

| Type of Occupancy | Portion of Lighting Load to Which Demand Factor Applies | Demand Factor (%) |
|---|---|---|
| Dwelling units | First 3,000 or less | 100 |
|  | From 3,001 to 120,000 | 35 |
|  | Remainder over 120,000 | 25 |

Notice that no reduction is permitted for the first 3,000 volt-amperes. In the two categories above that level, the demand factors are very generous, so this little table can make a very big difference in the size of the service.

# CHAPTER 4

# All About Electrical Services

The *electrical service* is variously defined to include all wiring from the utility point of connection to either the input terminals of the main disconnect/overcurrent device or to the entrance panel including its contents, all internal wiring and breakers. We'll use this latter definition only because it allows us to add a few comments about the entrance panel.

## Where Is the Point of Connection?

As far as the point of connection is concerned, that is for the utility to define. In most cases, it is where the utility workers actually connect their conductors prior to powering up the permanent service. This varies depending on whether the service is aerial or underground. If it is aerial, the point of connection is about 16 inches upstream from the weather head, where the utility crimps its triplex conductors to the customer-supplied service-entrance conductors, as shown in Figure 4-1. If it is an underground service, the point of connection is at the input lugs of the meter socket. The reason for this is that the underground cable is a single unspliced run from the meter to the transformer. It would not be good to have the home crafter-electrician (or even a professional electrician without high-voltage training) climbing the pole to make these connections.

As mentioned previously, the utility may require a licensed electrician to build the service and get it ready to be powered up. It may be possible for the home crafter-electrician to do the work, provided that a licensed electrician is willing to

**Figure 4-1**    For this aerial masthead service, the point of connection consists of the crimped splices upstream from the weather head. (The cable TV and telephone cables attached to the service mast are a Code violation.)

take responsibility for the installation with some degree of oversight. Regardless of who does the work, the home crafter-electrician should understand what is involved if for no other reason than because that provides perspective on the rest of the installation.

## Designing the Service

First, it must be emphasized that the utility should be contacted during the planning stage. If you build the service without utility consultation, it is possible that the installation will not be satisfactory from the utility's point of view, so expensive rework will be necessary before the utility will connect.

Most utilities have a detailed book of specifications that they give to electricians. This book has diagrams with explanatory wording on every conceivable service type and configuration, including labor and materials to be supplied by the customer.

The utility will have an engineer whose job is to view jobs prior to construction. It is not unusual for this individual to make more than one visit to the site to

**FIGURE 4-2**   The utility will have definite ideas regarding meter location. It is central to the design of the service.

work out details. Be certain that there is agreement on every facet of the installation—meter location including height, grounding, wire size, weather-head location, materials to be furnished by the utility, and so on (Figure 4-2).

## Building the Temporary Service

Almost all new construction projects and many remodeling projects require a temporary service, as shown in Figure 4-3. The purpose is to provide workers with electrical power needed to build the foundation and to do whatever work is required to construct the building to the point where it is possible to have an energized entrance panel inside. This usually involves having a roof on the building so that there will not be water damage to the electrical equipment. The temporary service is generally aerial, even if the permanent service is to be underground. The temporary meter and entrance panel are mounted on a wooden pole. The main object is to get the weather head high enough above grade so that when the utility connects its triplex about 16 inches upstream from the weather head, there will be sufficient ground clearance at every point along the line to the transformer. The temporary pole is usually 16 feet long. It may be 3 feet in the ground with 13 feet above grade. This should be high enough to get 12-foot ground clearance for the line from the temporary pole to the final utility power pole because the utility-owned triplex rises sharply as it runs toward the transformer. But this is not always the case. There may be a knoll or rise in the grade along the path, so it may be necessary to use a longer pole, make an extension at the top, or set it farther out of the ground. If this latter alternative is chosen, the pole should be braced to the ground, paying particular attention to the fact that the top of the temporary pole will try to move toward the utility pole because of the weight of the triplex, especially in winter, when there may be additional ice loading. Where it is impossible to maintain ground clearance, an alternative is to consider relocating the temporary service. All this should be worked out in advance with the utility representative.

Do not attach the temporary service equipment to the pole prior to setting it, and remove it prior to taking the pole down. Otherwise, you are asking for trou-

FIGURE 4-3    A temporary service. The two GFCI circuits need to be connected, and then it will be time to call the utility.

ble. The temporary pole is generally 6- × 6-inch pressure-treated lumber, but a segment of old telephone or power pole should be acceptable to the utility and electrical inspector.

Most electricians have two or three temporary services on hand so that separate jobs can be powered up simultaneously. They rent out these temporary services, and this may be the way to go because it will be cheaper than building a new temp, so you should price out both options. A rented temporary service will not include the pole.

To build a temporary service, proceed as follows: the backing board should be 2- × 12-inch pressure-treated lumber. Gone are the days of plywood, which gets ragged and delaminates when the edges are exposed to weather. It is best to assemble the temporary electrical equipment first before cutting the backing board so that it will be long enough. There should be room at the top to secure the service-entrance cable to avoid strain at the meter, particularly when transporting the temporary service between jobs. Also, be sure that there is plenty of room at the bottom for receptacles.

To assemble the temporary service, start with the meter socket. This is an expensive item, but sometimes the utility will donate a new or used one, complete with weatherproof hub. Be sure that the meter socket is the aerial type, not for underground services, and that it is a 100-ampere service instead of 200-ampere service unless it is an unusually large construction job.

Assemble the service equipment except for the service entrance cable, and mount it to the backing board prior to attaching the backing board to the temporary pole. The temporary service will have a more professional appearance if you make two diagonal 1-inch cuts at the upper corners of the backing board. At the top of the meter socket, you need a hub with waterproof connector, which has to be bought separately. Without this hardware, water will leak in and spoil the electrical connections. Assemble the breaker box to the meter socket. If the backsets of the two knockouts are the same, you can use a short length of metal raceway. If they are different, an offset nipple is needed.

## Bonding the Raceway

Because it is a service, this piece of metal raceway has to be specially bonded, beyond two locknuts at the meter socket end. For this redundant bonding, the inside locknut has to have a grounding lug, as shown in Figure 4-4. Take off this lug in order to tighten the locknut; then put it back on. The locknut has to be tightened sufficiently to dig into the paint on the inside of the box to ensure ground continuity. With this arrangement, where there is not a threaded boss, you

FIGURE 4-4   A bonding bushing with a ground lug connected to the grounding conductor is required for redundant bonding. It is usually located in the meter socket enclosure.

must have an insulating bushing on the end of the threaded raceway, and you have to remember this before making up the wire terminations.

## Ground-Fault Circuit Interrupters, Of Course

Assemble two outdoor cast-aluminum receptacle enclosures, commonly called *bell boxes*. You will need ground-fault circuit interrupter (GFCI) covers that are rated for use outdoors when cords are plugged into them to prevent water damage to the GFCIs. Use threaded offset nipples coming out of the bottom of the breaker box and into the top of the receptacle enclosures. Bonding locknuts are not required because these raceways are downstream from the overcurrent protection and are technically not part of the service.

Now that the four boxes are assembled, we are ready to mount them to the backing board. Use galvanized screws, and arrange the boxes precisely so that everything will be straight and plumb.

The next task is to wire the boxes, except for the service-entrance cable, which is hooked up after the backing board is attached to the temporary pole. First, connect the ground-electrode conductor at the meter socket. Most electricians provide a 2-foot whip outside the box so that between jobs they don't have to deal with the long piece. Use 6 American Wire Gauge (AWG) bare or insulated stranded copper wire. Thread an end through the miniature knockout on the bottom of the meter socket. No connector or sealing is required here because water should drain freely in the event that it finds its way into the enclosure. Bring the grounding-electrode conductor through the grounding-bushing lug and onto the meter-socket ground lug. Tighten these connections thoroughly so that if there is heavy fault current, they won't arc.

Next, wire the two bottom phase lugs in the meter socket. Then go through the offset nipples to the main breaker input lugs. Use Type THHN copper conductors of the required ampacity. You can use larger wire if it will fit in the main lugs.

Then insert two single-pole 20-amp breakers into opposite sides of the breaker box. Be sure that they are on opposite phase bus bars so as to balance the load. Using Type THHN 12 AWG black or red copper conductors, wire the breakers in this small service panel to the two GFCIs. Using white Type THHN 12 AWG conductors, wire the neutrals. Use bare or green wire of the same type to connect the equipment-grounding conductors. Check that the main bonding jumper is in place because this is a service panel.

Mount the backing board onto the temporary pole. Use galvanized lag screws so that it can be removed easily when the temporary service is taken down. Drive a ground rod nearby. Use a split-bolt connector to add enough 6 AWG ground wire to reach the ground rod. The ground wire should be stapled lightly to the temporary pole and then buried over to the ground rod. Connect this wire using a ground clamp.

At the top of the meter socket, the hub and connector with a watertight rubber seal should be in place. For the cable, use Type SE (service entrance), which is available in copper or aluminum. Copper is better, but everyone uses aluminum in this application because it is much less expensive. If you are using aluminum, choose 2 AWG (not 2/0 AWG) for a 100-amp service. For copper, the next smaller size, 4 AWG, is permitted because it has the same ampacity. Type SE is known as *concentric cable* because there are two ungrounded conductors inside a woven bare grounded conductor that is the neutral. This cable is quite safe because the grounded neutral is similar to a metal raceway. For service-entrance conductors, this extra protection is essential because the only overcurrent protection, at the utility transformer, is at a very high level. Even with a direct phase-to-ground short circuit, it won't trip out.

## Make It Watertight

Run the Type SE cable through the watertight connector, leaving a long-enough end. Using cable fasteners, secure the cable to the backing board and up the pole to the top. Then go back and tighten the watertight connector. This compresses the rubber and makes it grip the cable so that water will not run down inside. Just to be sure, apply a ring of silicone caulk at this connector.

At the top of the pole, install a weather head, fastening it to the pole using a removable lag screw. Be sure to run the neutral through the middle hole and the two hot phase conductors through the outer holes. To do this, slit the outer jacket and remove the scrap and the plastic strip. Separate the woven neutral from around the two ungrounded conductors, and twist the strands to make a single well-formed wire. Leave about 16 inches of each conductor including the neutral beyond the weather head so that the utility has plenty of wire to work with. The utility worker will form the wires to make a drip loop, strip the ends, and crimp them to the triplex to make an electrical connection. The utility worker also will provide a strain relief, lagged into the pole, so that the triplex does not pull on the weather head.

Back at the meter socket, inside the enclosure, slit the Type SE outer jacket, unbraid the neutral so that it no longer surrounds the ungrounded conductors, and twist it to form a single conductor. Terminate the grounded conductor at the neutral lug and the two hot ungrounded phase conductors at the two phase lugs. Form the wires inside the enclosure to make gentle loops as opposed to straight bars going to the lugs so that thermal motion doesn't compromise the connections. To make the phase and neutral connections, it is best to remove the lug screws with contacting plates, lay the stripped ends in place, and then replace the hardware. All service terminations, especially if the conductors are aluminum, should be quite tight.

## Special Techniques for Aluminum Terminations

Aluminum terminations have the disadvantage that they are prone to failure if not done correctly. For this and other reasons (such as conduit fill), in short runs where cost is not decisive, copper is used. If an aluminum connection is made, over a period of time, the aluminum seems to flow away from the mating metal surface. Eventually, the electrical connection will begin to heat up, especially if there is heavy current. This sets the stage for corrosion, less conductivity, more heat, more corrosion, and so on until there is at best an outage and ruined lugs and wire ends and at worst an electrical fire.

The remedy is to use corrosion inhibitor on all aluminum terminations. Follow the directions on the container. Wire brush the stripped ends and inside the lugs. Be sure that there are no insulation fragments that could get caught between mating surfaces. Beyond the electrical connection, there should be at least ¼ inch of wire so that the lug does not tend to expel the conductor when it is tightened. Be sure to tighten the connections correctly. Use the manufacturer's specifications and a torque wrench.

Check all terminals with your ohmmeter. With no connected loads, there should be no continuity between the two legs or between either leg and ground. However, there should be continuity between the equipment-grounding conductor and the neutral and all metal enclosures and the grounding-electrode conductor.

Replace the cover on the meter socket. There should be a cardboard closure pad that goes in the meter-socket opening. The utility may give you a reusable plastic shield that works quite well. Review the entire installation. If everything looks good, call the utility to have it heat up the temporary service.

Soon it will be time to build the permanent service, switch over the power, and decommission the temp. At most, there will be no more than a few minutes that the building is without power.

Electrically, the permanent service, shown in Figure 4-5, is the same as the temporary service. The physical layout is different because it is attached to the

**FIGURE 4-5**  This meter socket is in back-to-back configuration with the entrance panel on the inside of the house. The polyvinyl chloride (PVC) conduit coming out of the bottom of the meter socket contains the grounding-electrode conductor.

building. There is a great variety in types of permanent services, with many kinds of hardware available. You have some freedom to improvise, within limits set by the *National Electrical Code* (NEC), the utility, and the local jurisdiction. Throughout, remember that there is tremendous available fault current without the usual overcurrent protection.

## Building a Service

Learning to build a service is greatly facilitated by the fact that you can drive around any suburban neighborhood and view many types of services. A good part of each one is hanging on the outside of the building, where it may be scrutinized by the inquisitive worker without ever setting foot on the property. A digital camera will permit assembling a collection that may be put into a folder on a computer. Together with the utility book of specs, the *NEC*, and suggestions in this book, any type of service should be doable.

There are two subcategories of the permanent service—aerial and underground. Inside the house, they are identical, but the outside physical layouts are different. The underground service is more expensive, but it has some advantages, aesthetic and otherwise. Because there is no overhead service drop to impede the view, the property has a less cluttered appearance, and in the end, the real estate becomes a more valuable asset. Without a weather head and service-entrance cable attached to the exterior wall, the building, often with a gable end facing the road, has a more stately appearance. Where the aerial cable crosses the yard, the utility has an easement whose width may be substantial, and this can affect use of the property. Local septic regulations often prohibit a tank and leach field within this easement, and for a small lot, this may limit options. With an underground service, it is permissible to wind around to an extent as long as the bends are large-radius sweeps. If the home crafter-electrician does the labor and reasonable backhoe arrangements are possible, an underground service is feasible, and it will be much appreciated in years to come.

We shall look at both types of service from the construction point of view, beginning with the aerial service. In the course of the initial meeting with the utility engineer, a meter location will have been determined, perhaps with a range of options for the property owner. The utility wants the meter to be in an easy location to read, preferably without the reader having to leave his or her vehicle. This is true even if it is a "smart" meter because you never know when the remote communication will become problematic. The utility will want the meter protected from damage, on the gable end rather than the eave side, where there is a possibility of falling ice, and located so as to accommodate straight-line connection to the transformer. Additionally, the utility will want an eye-level height for ease of reading, which translates to a height of 5 feet above the grade. Don't forget to allow for any deck or addition that may be built. Other than these requirements, the meter location is essentially up to the homeowner.

## Masthead Service Construction

Materials for the masthead service are a little more expensive, and more labor is involved. In this configuration, the meter is located on the eave side of the house, which is necessary for mansard or double-hip-roof construction, where there is no gable end. This location is also used where the gable end is not high enough to provide minimum ground clearance for the service drop.

Meter placement is affected by a strong *NEC* mandate that applies to the inside portion of the building. It is stated that the main disconnect must be located closest to the point where the service-entrance conductors enter the building. The mandate calls for some interpretation because, if nothing else, there will be the thickness of the wall to consider. Nevertheless, a back-to-back pattern is invariably acceptable and is in fact the best. This construction involves removing large knockouts on the backs of both entrance panel and meter socket, inserting a straight threaded conduit whose length is equal to slightly more than the thickness of the wall, and passing the service-entrance conductors through that short raceway. Exposure to damage for these conductors is minimal.

## Service Variations

For other setups, the precise permitted separation will be a matter of interpretation for the electrical inspector. As much as 12 feet is sometimes mentioned as a rule of thumb. The installation will be seen in a more favorable light if the conductors are run in a metal raceway. In a problem site, contact the electrical inspector prior to construction.

Often in slab construction, where there is no basement, the back-to-back configuration works well, even though the entrance panel may have to be a little lower than the optimum eye level. This is not a Code violation.

When there is a full basement, there may be a problem in locating the entrance panel because of various factors, such as the location of a fuel tank or water piping. Also, the entrance panel, because it contains overcurrent devices, may not be located in a bathroom, clothes closet, on a stairway, or anywhere that there is not sufficient working space or dedicated space above.

In some buildings, any of these factors may severely limit placement of the entrance panel. One way to work around the Code mandate mentioned earlier is to have a main disconnect that is separate from the entrance panel. Then, because they are overcurrent protected, the conductors to the entrance panel become a feeder, which can be placed anywhere on the property. In this case, you can use what is called a *main-lugs-only breaker box*. It has no main breaker and is no longer an entrance panel, becoming what is known as a *load center*. Of course, if you already have a box containing a main breaker, there is no harm in that. In such a construction, we must stress that there is to be no main bonding jumper in this box. Instead, it is located in the main disconnect enclosure. The feeder, then, has four wires, including the green equipment-grounding conductor.

The remote main disconnect makes for a more expensive installation and is less convenient, but for some sites, it is necessary to comply with the *NEC* mandate mentioned earlier. The main disconnect is available as an outdoor unit, and this is necessary in some cases.

In view of these considerations, you have to determine good entrance panel, main disconnect, and meter locations before proceeding with the service. In some remodeling jobs, the temporary meter occupies the space where the permanent meter has to go. With a little advance planning (leaving extra slack in the service entrance and grounding-electrode conductors), it is possible to swing the temporary meter off to the side so that the permanent meter socket can be mounted in place and wired.

To do a new gable-cable aerial service, if it is a back-to-back installation, first use a hole saw to drill a hole in the sheathing. This hole should be about ⅛ inch larger than the outside diameter of the threaded conduit that is going to attach to the back of the meter socket and the back of the entrance panel. So begin by checking the sizes of these two knockouts. They may not be the same.

These knockouts may be offset from the center in one or both boxes, so the whole thing has to be laid out paying attention to the locations of any studs that may block the hole, how the entrance panel door is to swing, whether backing boards are to be installed inside and out, and similar considerations. If the siding is not yet on the outside wall, you may want to make a pine backing board for the meter socket. It should be about 1 inch bigger all around. On the inside, is the entrance panel flush-mounted between studs or surface-mounted on the finish wall? You have to think about numerous issues, such as whether it will be possible to take out the knockouts and add connectors and wiring in the future if the entrance panel is close to studs. One solution, in a finished living space, is to provide a backing board for a surface-mounted breaker box and then (later) build a cabinet around it with a good-finish access door.

The following discussion presupposes that you have read the section of this book on temporary services, especially the parts about bonding service equipment to metal raceways and use of corrosion inhibitor with aluminum conductors. With the hole drilled and the boxes set in place, measure the length of the threaded conduit nipple that you will need. This is a difficult measurement to make because you have to have enough threaded ends at both boxes to accommodate the bonding locknut at the meter socket, the locknut at the entrance panel, and the insulating bushing. It is recommended that you do your best to get an accurate measurement and then obtain the next sizes longer and shorter as well.

One of the advantages of the back-to-back arrangement is that the two boxes hold one another tightly to the wall even without the mounting screws, which nevertheless should be used to keep the boxes from turning. The other advantage is that there is less clutter inside and out. The service-entrance conductors are invisible. Before installing them, it is best to complete the grounding.

## Grounding-Electrode System

A number of grounding-electrode options are available. Some of them are

- **Metal underground water pipe.** This must be in direct contact with the earth for at least 10 feet. The problem with this type of grounding electrode is that increasingly polyvinyl chloride (PVC) is being used. Even with a metal pipe, you never know whether plastic has been spliced in, rendering the water pipe grounding electrode ineffective.
- **Metal frame of the building where at least one structural member is in direct contact with the earth.** This choice is not available in wood-frame residential buildings.
- **Concrete-encased electrode.** This is to be at least 20 feet of either one or more bare or zinc-galvanized or other electrically conductive coated-steel rebar no less than ½ inch in diameter. If there are multiple pieces, they may be connected by the steel tie wires. Metal segments must be encased in 2 inches of concrete. This is called a *ufer*, and it is highly effective, but it won't be available unless provisions were made at the time the concrete was poured.
- **Ground ring encircling the building.** This must consist of 20 feet of bare copper conductor not smaller than 2 AWG.

- **Ground plate with 2 square feet of surface exposed to exterior soil.** This is generally copper, at least 0.06 inches thick.
- **Ground rods 8 feet long.** These are generally made of steel, copper coated, or galvanized.

In most new construction, ground rods are used. The *NEC* specifies that ground resistance must be measured, and the resistance of the ground rod to ground must be not more than 25 ohms. Unfortunately, ground resistance cannot be measured directly with an ohmmeter because this would presuppose the existence of a grounding electrode for reference having a ground resistance of substantially zero, in which case there would be no point in going further. Expensive equipment with elaborate procedures will measure ground resistance, but the *NEC* provides an alternative—an exception states that if a second ground rod is installed, it is not necessary to take the resistance measurement.

Most new residential construction makes use of two ground rods. When this is done, the rods must be at least 6 feet apart so that overlapping ground potentials do not reduce the overall effectiveness. If bedrock is encountered, it is permissible to drive the round rods at up to a 45-degree angle or to bury them in a trench with 24 inches of cover material. Under no circumstances should a ground rod be cut short.

The grounding-electrode conductor must be attached by means of a ground-rod clamp, never a hose clamp or improvised means. The grounding-electrode conductor should be run first to the near rod and then to the far one. This wire should be buried to prevent damage and for increased grounding. Both rods should be driven below grade so that the whole system is out of sight.

From the meter socket to below grade, the grounding-electrode conductor must be in a raceway. If the raceway is metal, at the bottom end it must be bonded to the grounding system. For this reason, most electricians use PVC conduit. This piece must have a two-point offset bend, as shown in Figure 4-6, so that beginning about 6 inches below the meter socket, it points toward the wall and then turns again at the same angle but in the opposite direction so that it runs along the wall down to a level several inches below grade. This is always the way to run raceway offset bends so that the raceway hugs the wall or ceiling as opposed to flying through the air.

In small diameters, PVC conduit can be bent using the next-size-larger electrical metallic tubing EMT bender. Bend the pipe twice as far as the desired angle, and it will spring back as needed. The other method is to use heat to soften the PVC conduit sufficiently so that it can be easily bent by hand. Don't use a propane torch because it will scorch the outside without heating the inside. When you try

**Figure 4-6**   PVC conduit for the grounding-electrode conductor can be bent by hand after heating it over a charcoal grill. (Practice on scraps first!)

to make the bend, the pipe will kink, substantially reducing the inside diameter. There is a professional electrician's tool that resembles an electric blanket; it fastens around the pipe and heats it evenly, doing a nice, clean job. The problem is that the tool is very expensive, and the element is prone to burning out. There are some PVC heat benders that work off an automotive exhaust, but who knows what the petroleum residue will do to the wire in the long run? I have had good results holding the pipe about 16 inches above a charcoal grill, turning the pipe and moving it lengthwise to get the heat where you want the pipe to bend. Practice on scraps.

The PVC pipe is cemented to a threaded adapter and fastened through the appropriate knockout using a locknut. Secure it to the wall below the offset using conduit fasteners. Run the grounding electrode through the PVC conduit, into the meter socket, and through the bonding locknut, terminating it at the grounding lug.

Recent *NEC* editions require an intersystem-bonding terminal. Using a split-bolt connector, tap a segment of 6 AWG bare stranded ground wire onto the grounding electrode, and bring it out through the miniature knockout with no connector. Run the ground wire a short distance with no bends, and attach an intersystem-bonding terminal, which is fastened to the outside wall. It has a removable cover, and the purpose is to allow other trade workers such as telephone and satellite dish installers to bond their ground wires to the building's overall grounding system. It is a long established principle that all grounding systems of a building must be bonded together because this prevents dangerous voltage potentials between them. Some people feel that this bonding could call lightning into the building, but this is not the case.

Wire the meter socket to the entrance panel. In a back-to-back configuration, Type THHN copper works well. For 6 AWG and larger wire, all three conductors are permitted to have black insulation because it is not feasible to carry around separate reels of each color. There is no identity between the two hot legs, so it doesn't matter if they cross over between meter socket and entrance panel. The neutral, however, absolutely must not be confused with a hot phase wire or there will be a terrible arc flash when the utility worker attempts to insert the meter. The black neutral must be reidentified at both ends. The ends can be painted, but a better method is to use phase tape. This resembles electrical tape but is available in several colors, including red, white, and green. Wrap the correct color close to the stripped end of the conductor. The reidentifying means must completely encircle the conductor. Some electricians make three rings, but one is sufficient.

Reidentify both ends of the neutral using white phase tape. Then hook them up. Afterwards, run and terminate the hot phase legs. Check the installation with your ohmmeter to make sure that there are no shorts.

Some utilities furnish the cable that goes up the side of the building, and they also furnish the weather head. Most do not, so this segment is left to the owner.

Wires from the meter output lugs to the main breaker and from the meter input to the weather head are both the same size unless one is copper and the other is aluminum. They may be different types, for example, Type THHN separate conductors through a raceway from the meter socket to the entrance panel and concentric cable (Type SE) up the side of the building. They are different types of wire but are sized using the tables in *NEC* Chapter 3.

## Service Sizes

For residential work, the most common service sizes are 100 and 200 amperes. A 100-ampere service calls for 4 AWG copper or 2 AWG aluminum. A 200-ampere service requires 2/0 copper or 4/0 aluminum. These are the sizes to remember. Some electricians like to use an intermediate size such as a 150-ampere service.

In all but the largest houses, a 200-ampere service is more than adequate, even with electric heat and a workshop with an arc welder and power tools. In a small cottage, a 100-ampere service is often installed, only to find as years pass that electrical use increases and a service upgrade is needed. This is a costly process because you have to replace not only the entrance panel but also the service-entrance conductors and weather head as well. And it is likely that the branch-circuit conductors will have to be extended to reach the breaker and grounding terminals inside the larger box.

It is interesting to note that a 200-ampere service is actually more energy efficient than a 100-ampere service. Devices such as breakers, wire nuts, and conductors are not considered to be consumers of electricity, but with so many gathered in a small volume, together they emit some heat, which makes the meter turn. A 100-ampere entrance panel, heavily loaded, will feel warm to the touch. A 200-ampere entrance panel will run cooler because all the devices, terminations, and conductors are larger with less $I^2R$ loss.

A variation of the gable-cable service is the gable-conduit arrangement. This makes a neat job, and the service-entrance conductors are well protected. Raceways can be PVC or EMT. Conductors are Type THHN, usually aluminum because this is a long run. Be sure to use the weather head designed to go with the raceway. It clamps onto the pipe rather than gripping the concentric cable.

Both gable-cable and gable-conduit types do not necessarily have to go to the peak of the house. Just run them high enough to get the required ground clearance. Sometimes the building has a low roof, and it is not possible, using the services described earlier, to achieve the proper ground clearance for the triplex service drop. This also has to do with the terrain between the building and the utility pole. A small knoll under the cable right where there is maximum sag can reduce ground clearance. When it comes to ground clearance, there are obviously important safety concerns.

Where the building is not high enough to accommodate a conventional service, the masthead type is the way to go. A metal raceway comes out of the top of the meter socket, penetrates the soffit and the roof, and rises to a height sufficient so that ground clearance is achieved. A masthead service should not be considered if it is not necessary for the service drop ground clearance or to locate the meter on a gable side. It is more expensive in materials, and the soffit hole and roof penetration, including the leakproof roofing job, mean a lot of extra work.

The soffit hole must be accurately positioned with respect to the meter location, taking into consideration the backset of the meter hub. If this hole is at all out of position, the raceway above and below the roof will not be vertical. A good way to proceed is to temporarily fasten the meter socket in place at the usual 5 feet above grade. Then use a plumb bob to find the center of the hole-saw cut in the soffit. When you get a little further along, you can shift the meter location a slight amount if necessary.

For the metal raceway, rigid metal conduit (RMC) is used, not water pipe, which would be a Code violation. The outside diameter and threads are identical, but RMC has a smoother inner surface so that the conductors are not damaged, and they are easier to pull. Ordinarily, a 10-foot length of 2-inch RMC will suffice. If not, the RMC is available in 20-foot lengths, which can be cut down as needed. Also, couplings can be obtained. Where the raceway goes through the roofing, a rubber boot or flashing will ensure a leakproof penetration.

The *NEC* states that no foreign wiring or other paraphernalia is to be attached to the masthead and certainly not run down the raceway along with service-entrance conductors. Very often telephone installers anchor their drop onto the electrical service masthead, and this is a definite Code violation.

At the top, use a compatible weather head, as shown in Figure 4-7. Fish in Type THHN aluminum conductors sized as for any other service. The utility will crimp onto the 16-inch conductor ends that you leave sticking out of the weather

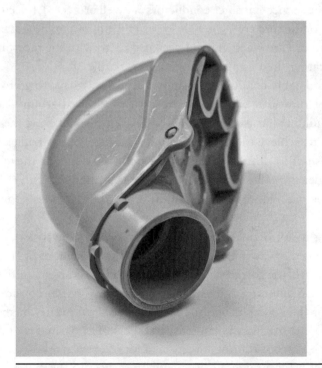

**Figure 4-7**   A weather head, available in metal or PVC, is used at the top end of an aerial service, cable gable, or masthead.

head. Grounding and wiring into the entrance panel or main disconnect are identical to those of other services.

## Underground Services

Underground services are somewhat more expensive to build, but they contribute to making an upscale building. There is no service drop to impede the view and no service-entrance cable or weather head to clutter the finish wall. With a back-to-back meter and entrance-panel hookup, the underground service makes for a simple and elegant final product, and it will enhance the value of your real estate.

Generally speaking, a backhoe is needed to dig the trench from the utility pole to the meter location. The question that always arises is, how deep does an electrical line have to be buried? To answer this question, refer to *NEC* Table 300.5.

Assuming that it is a back-to-back configuration, the underground hookup is quite simple, although more labor is involved than for an aerial service. The consultation with the utility representative will nail down the details. Generally, the customer digs the trench and furnishes the underground run of conduit, cemented with a pull rope in place.

The conduit should be installed as a complete system, hooked up to the meter socket, and the trench backfilled and graded. The utility usually furnishes the service lateral conductors. The utility pulls them through the conduit and makes the terminations.

The telephone line should be buried in the same trench with maximum separation from the power line. Use 2-inch PVC with sweeps and a pull rope. Leave stubs 1 foot above grade at both ends. The telephone company, subject to prior consultation, will pull in its line, put an interface box on the wall, and make all terminations, including bonding to your intersystem-bonding terminal.

If the soil is at all rocky, the conduit should be bedded in screened sand to a height of 6 inches above the conduit. Then another 6 inches of native fill is added by hand to make sure that no big rocks damage the conduit. If you place the conduit against the edge of the trench and machine backfill from that side, there is less exposure to damage.

The service lateral remains the property of the utility, and if there is a problem in the future (such as lightning burnout), it is the responsibility of the utility to make the repair. Because the service lateral is in conduit, it can be replaced even if the ground is frozen with no digging required. For this reason, even if the conductors are rated for direct burial, they are always put in raceway. PVC is the conduit of choice for almost all underground work. In consultation with the utility, you probably will use steel 90-degree sweeps and expansion joints to allow for ground movement if subject to freezing.

At the building, place the sweep so that the stub will come up perfectly plumb with the meter-socket knockout. You can finish this prior to drilling through the building for the back-to-back stub so that the meter socket location can be adjusted laterally to make for a straight riser.

The expansion joint should be put in with the outer part of the slider at the top so that it will shed water. Position the expansion joint midway between maximum and minimum length. Grounding and wiring to the entrance panel are the same as for an aerial service.

At the utility pole, details should be worked out in advance. Typically, to a height of 8 feet above finish grade, Schedule 80 PVC is used. (All other PVC is Schedule 40.) Because the sweep contributes some rise, a 10-foot length of Schedule 80 PVC ordinarily will do. Most utilities want to see this piece installed by the electrician. Then the utility furnishes the rest of the run, using Schedule 40 PVC up to the transformer, where there is a weather head and a strain relief with a drip loop.

Regardless of the type of service, the back-to-back arrangement of meter socket and entrance panel or main disconnect is preferable. There is less clutter inside and outside the building, and minimal conduit and wire are needed to get into the building. Moreover, both boxes are held very firmly in place and will never work loose.

Sometimes, usually because of the vertical layout of the building, a back-to-back configuration is not possible. In such cases, we have already mentioned the need to install a separate disconnect if the indoor portion of the service-entrance conductors are of any significant length.

In a non-back-to-back installation, the service-entrance conductors can be run as Type SE concentric cable or in a raceway. If cable, it is best to come out of the meter socket at the bottom. Such an arrangement is preferable because water infiltration is not an issue. It will drain straight out of the bottom.

Where the cable enters the building, a small piece of hardware known as a *sill plate* is used. It is sized to fit the cable. The cable should enter the sill plate from below so as to shed water. Fill around the cable with silicone caulk.

These conductors also can be run in a raceway. PVC, RMC, or EMT can be used. PVC conduit (gray UL listed, never white PVC water pipe) should not be used in long horizontal runs on the outside of a building because thermal changes will make it sag and buckle. EMT, if used outdoors, must have compression fittings. Set-screw fittings would allow water to enter, and they are used indoors only.

Ordinary 90-degree bends, as found in water systems, are not used in electrical raceways. Instead, pull boxes are needed, as shown in Figure 4-8. These fittings have removable plates, making it much easier to pull the conductors.

FIGURE **4-8**   An LB, available in metal or PVC, is used where an underground service lateral enters the building.

# CHAPTER 5

# The *National Electrical Code*: Fundamental Requirements for Residential Work

Professional electricians are guided in every detail of their work by the *National Electrical Code* (NEC). The home crafter-electrician likewise must follow the requirements contained in this extensive document. The object is to create an electrical installation that is not a hazard to the worker(s) during construction and also will not endanger the end users. If you are doing electrical work, regardless of size and even if it is limited to your own home, it is essential to possess a copy of the *NEC* and to refer to it at each stage of construction.

## Understanding the *NEC*

Because the *NEC* is lengthy and complex, it is sometimes a challenge to find the information you need in a timely fashion. You don't want to be Code hopping, just flipping pages in hopes of finding an answer. You need to have a system for navigating throughout this 600-page document (Figure 5-1). The key is to understand its structure. It is useless to attempt to memorize all the Code mandates or even a fraction of them. If you could count all the bits of information including data in tables, the number would be overwhelming. So how do you proceed?

A certain amount of memorization is needed, but it is not an impossible task. To begin, learn the titles of the nine chapters in their correct order. This will go far in helping you to visualize the overall Code structure.

The *NEC* is revised every three years, and the changes are invariably numerous and have a great impact on the way that work is to be done. Although some

**Figure 5-1**    The *NEC Handbook*, containing the complete Code text plus commentaries, photographs, and diagrams.

jurisdictions are slow to adopt the current *NEC* edition, others jump right on the wagon in a timely fashion. The timing can be clarified by contacting your local zoning board. In the discussions that follow, we'll focus on the 2014 *NEC*. Except for a few isolated instances, this will work. Many changes in the 2014 *NEC* are not applicable to residential locations, and others are minor alterations of syntax or wording that do not have to concern us.

## 2014 *NEC* Changes

Here are the principal 2014 *NEC* changes that apply to residential work and are essential for the home crafter-electrician:

- **Section 110.26(E)(2), "Dedicated Equipment Space. Outdoor."** The dedicated space for electrical equipment is equal to the width and depth of the equipment, and it extends from the grade to a height of 6 feet above the equipment. No piping or other equipment that is not part of the electrical installation is permitted to be located in this zone. Gas piping; water piping; refrigeration lines; and phone, cable, and satellite equipment may not occupy the dedicated space reserved for a meter socket, entrance panel, or other electrical equipment found in a residence, and this requirement applies indoors and out as well.

- **Section 210.8(A)(7), "Ground-Fault Circuit-Interrupter Protection for Personnel. Dwelling Units."** Any 125-volt, single-phase, 15- or 20-ampere receptacle within 6 feet of the outside edge of a sink in a dwelling unit including kitchens is required to have ground-fault circuit-interrupter (GFCI) protection as provided by the GFCI receptacle shown in Figure 5-2.
- **Section 210.8(A)(9), "Ground-Fault Circuit-Interrupter Protection for Personnel. Dwelling Units. Bathtubs or Shower Stalls."** All 125-volt, single-phase, 15- and 20-ampere receptacle outlets in bathrooms and those outside the bathroom that are within 6 feet of the edge of a bathtub or shower stall are to be GFCI protected.
- **Section 210.8(A)(10), "Ground-Fault Protection for Personnel. Dwelling Units. Laundry Areas."** All receptacle outlets in dwelling-unit laundry areas require GFCI protection, including the outlet for the washing machine. These GFCI receptacles in the laundry area must be readily accessible.
- **Section 210.8(D), "GFCI Protection. Kitchen Dishwasher Branch Circuit."** In dwelling units, kitchen dishwasher outlets are to have GFCI protection.

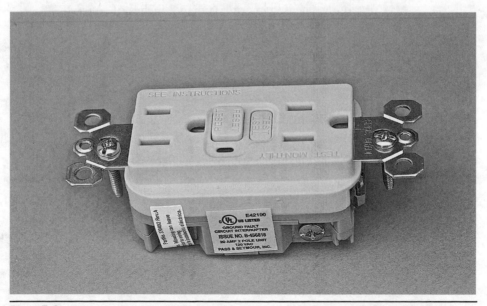

**FIGURE 5-2**  The receptacle-type GFCI device has feed-through capability so that conventional downstream receptacles can be protected.

## Arc-Fault Protection

- **Section 210.12(A), "Arc-Fault Circuit-Interrupter Protection. Dwelling Units."** Arc-fault circuit-interrupter (AFCI) protection has been extended to include kitchen and laundry areas. These areas have been added to the previous list, which included family rooms, dining rooms, living rooms, parlors, libraries, dens, bedrooms, sunrooms, recreation rooms, closets, hallways, and the like. In dwellings, AFCI protection is not required in bathrooms, garages, crawl spaces, attics, and outdoors.
- **Section 210.12(A)(1)–(16), "Arc-Fault Circuit-Interrupter Protection. Dwelling Units."** Besides simply installing a listed combination-type AFCI in the entrance panel or load center to protect the entire branch circuit, there are other methods that are Code compliant:
  - Install an outlet branch circuit–type AFCI receptacle as the first outlet on the branch circuit. The run between the circuit breaker and the first outlet must be rigid metal conduit (RMC), intermediate metal conduit (IMC), electrical metallic tubing (EMT), metal-clad cable (MC), or steel armored cable (AC), and the outlet and junction boxes must be steel.
  - Install an outlet branch circuit–type AFCI receptacle as the first outlet on the branch circuit, with the raceway from the circuit breaker to the first outlet encased in not less than 2 inches of concrete.

These are the traditional (2011 *NEC*) methods of achieving AFCI protection. Where an in-panel AFCI breaker (which is very expensive) is not used, equivalent

---

### Type MC Metal-Clad Cable

Metal-clad cable (Type MC), shown in Figure 5-3, is a factory assembly of one or more insulated circuit conductors with or without optical fiber members enclosed in an armor of interlocking metal tape or a smooth or corrugated metallic sheath. Uses permitted include services, feeders, and branch circuits for power, lighting, control, and signal circuits. Type MC cable may be used indoors or outdoors, exposed or concealed. The minimum bending radius for interlocked-type armor or corrugated sheath is seven times the external diameter of the metallic sheath. Sharper bends will cause kinking and damage to the conductors inside. Type MC cable is to be secured and supported by appropriate hardware at intervals not exceeding 6 feet and secured within 12 inches of every box, fitting, cabinet, or cable termination.

**Figure 5-3** Type NM cable, at left, has many uses, such as powering a hot-water heater or submersible well pump for a home.

protection in the form of raceway, concrete encasement, or length limitation is provided for the conductors that are not AFCI protected.

The 2014 *NEC* offers added options:

- Install a listed branch-circuit/feeder-type AFCI breaker and a listed outlet-type branch-circuit AFCI receptacle as the first outlet on the circuit. The first outlet box must be marked to show that it is the first outlet on the circuit.

---

**Type AC Armored Cable**

Armored cable (Type AC) is similar in appearance and use to Type MC cable. The principal difference is that it has an interior conductive metal strip in continuous contact with the armor for purposes of enhanced equipment grounding.

This is a moot point in residential work, where Type MC has an enclosed green equipment-grounding conductor, and redundant grounding (as in a health-care facility) is not needed. The supporting interval is 4½ feet.

---

- Install a listed supplemental arc-protection circuit breaker with a listed outlet branch circuit–type AFCI receptacle as the first outlet on the circuit, with conditions.

- Install a listed outlet branch circuit–type AFCI as the first outlet on the branch circuit in combination with a listed circuit breaker, with conditions.

The point of these alternatives is to save the expense of buying the breaker-type AFCI. After all, at a discount source, they still cost over $35, and at one for each branch circuit in a residence that requires AFCI protection, this adds substantially to the price of building a new home or large addition. However, how much will be saved unless the first receptacle is very close to the entrance panel?

## Some More 2014 *NEC* Changes

- **Section 210.17, "Electric Vehicle Charging Circuit."** Outlets installed for the purpose of charging an electric vehicle are to be on a separate dedicated circuit with no other outlets.
- **Section 210.52(E)(1), "Outdoor Dwelling-Unit Receptacles."** The outdoor receptacles, required at the front and back of each residence, no longer have to be accessible from grade level, so if one or both of them are accessible only from a deck, that will meet the requirement. In wiring a new home or addition, don't neglect to provide for the two outdoor receptacles, whether accessible from grade level or not.
- **Section 210.52(G), "Receptacle Outlets, Basements, Garages, and Dwelling Units."** In an attached garage or a detached garage with electric power, at least one receptacle outlet must be installed for each car space, and the branch circuit supplying these outlets cannot supply other outlets outside the garage. The underlying principle of these new requirements is that a garage that is attached to a residence must be connected to the premises wiring, whereas power for a detached garage is optional.

## Wet-Location Receptacles

- **Section 406.9(B)(1), "15- and 20-Ampere Receptacles in Wet Locations."** The 2014 *NEC* requires all 15- and 20-ampere receptacles in wet locations to have extraduty covers. The 2011 *NEC* required in-use covers, but they were found to break easily and not to perform their intended function of keeping water out of the receptacles when cords were plugged into them. The new extraduty covers are more durable and are now required.

- **Section 406.12, "Tamper-Resistant Receptacles."** Tamper-resistant receptacles are required for nonlocking 125-volt, 15- and 20-ampere receptacles in dwellings. They are not required under the following conditions:
  - Where receptacles are more than 5½ feet above the floor
  - Where receptacles are part of a luminaire (light fixture) or appliance
  - Where it is a single or duplex receptacle for two appliances located within a dedicated space and not easily moved
  - Where it is a nongrounding receptacle used as a replacement

The purpose of tamper-resistant receptacles is to protect naturally inquisitive children from inserting conductive objects into the receptacles, resulting in electric shock.

These are the principal 2014 *NEC* changes that apply to residential work. Besides being important to promote electrical safety, they should be incorporated, where relevant, into any electrical installation because the inspectors focus intently on these new requirements.

## Some Definitions

There are perennial definitions, existing language that reappears in each new Code edition with a word changed here and there but substantially the same. Here are the principal terms that pertain to residential work and are of greatest importance for the home crafter-electrician:

Throughout the *NEC* we encounter the terms *accessible, readily accessible,* and *concealed.* They are significant because, throughout the *NEC,* many electrical devices and types of equipment are required to be located in, permitted in, or prohibited from these locations, so first you need to know how to delineate them.

*Accessible,* as applied to equipment, means admitting close approach, not guarded by locked doors, elevation, or other effective means. *Accessible,* as applied to wiring methods, means capable of being removed or exposed without damaging the building structure or finish or not permanently closed in by the structure or finish of the building. *Readily accessible* means capable of being reached quickly for operation, renewal, or inspection without requiring those for whom ready access is necessary to use tools, to climb over or remove obstacles, or to resort to portable ladders and so forth.

This distinction, for the most part, is easy to understand, but gray areas are possible, and in such instances, the judgment is to be made by the electrical inspector. An example of a location that is accessible but not readily accessible is the space above a suspended ceiling. A switch adjacent to a room entry that controls a ceiling light must be readily accessible, and of course, this is a profound safety issue when, to mention one example, there is a fire and rescue workers have to check rooms.

All splices in electrical wiring must be accessible but need not be readily accessible. Junction boxes containing electrical splices are not to be concealed behind finish walls or ceilings. This is a long-standing *NEC* violation and interferes greatly with troubleshooting and repair. Such enclosures used to be called *blind boxes*, but the term is no longer used out of respect for the visually impaired.

A related concept is *concealed*, which means rendered inaccessible by the structure or finish of the building. Various types of wiring are required, permitted, or prohibited from being concealed, and to disregard one of these provisions is a definite Code violation. If it is caught by an inspector, expensive rework may be required, and if the problem is not detected, there could be a hazardous situation at some time in the future.

A *bathroom* is defined as an area including a basin with one or more of the following: a toilet, a urinal, a tub, a shower, a bidet, or similar plumbing fixtures. According to this definition, a simple washroom with only a sink and mirror would not be considered a bathroom, and therefore, the entrance panel could be located there.

Similarly, a *clothes closet* is defined as a nonhabitable room or space intended primarily for storage of garments and apparel. If there is a problem finding a location for the entrance panel, it may be possible to repurpose a clothes closet, but in that case, clothes and apparel would have to be kept out of the room or area. Also, dedicated space and clear working space would have to be verified.

A *kitchen* is defined as an area with a sink and permanent provisions for food preparation and cooking. The presence of a microwave oven in a room would not in itself constitute a kitchen, and this distinction would be decisive in determining whether to provide GFCIs and the two 20-ampere small-appliance circuits.

A very basic requirement for all electrical wiring, residential, commercial, or industrial, appears in Section 110.26, "Spaces about Electrical Equipment." It provides that access and working space are to be provided and maintained about all electrical equipment to permit ready and safe operation and maintenance, as shown in Table 5-1.

Two voltage categories, 0–150 and 151–600, appear in the left column. The table provides minimum distances for three conditions:

- **Condition 1.** Where exposed live parts are on one side of the working space and no live or grounded parts are on the other side of the working

TABLE 5-1  Minimum Clear Distances (Depths) for Various Conditions and Voltages to Ground

| Volts to Ground | Minimum Clear Distance Condition 1 (feet) | Minimum Clear Distance Condition 2 (feet) | Minimum Clear Distance Condition 3 (feet) |
|---|---|---|---|
| 0–150 | 3 | 3 | 3 |
| 151-600 | 3 | 3½ | 4 |

space or exposed live parts are on both sides of the working space that are effectively guarded by insulating materials.

- **Condition 2.** Where exposed live parts are on one side of the working space and grounded parts are on the other side of the working space. Concrete, brick, or tile walls are considered grounded.
- **Condition 3.** Where exposed live parts are on both sides of the working space.

Notice that Table 5-1 applies to the depth of the working space. The width of the working space is a different matter. It is the width of the equipment or 30 inches, whichever is greater. Most electrical equipment that falls into this category in residential settings (e.g., entrance panels and smaller disconnecting means) is less than 30 inches wide, so the required width of the working space is generally 30 inches. If two or more enclosures are mounted side by side, they may share this space. Moreover, a box need not be centered in the working space as long as the entire enclosure is within it.

The working-space requirement is entirely separate from the dedicated space above equipment, and both of these must be met, both indoors and outside. A Code violation is shown in Figure 5-4. The purpose of the dedicated space is to ensure that the equipment can be installed and will function properly, whereas the clear-working-space requirement focuses on worker safety. The height of the working space is to be the height of the equipment or 6½ feet above the grade, floor, or platform, whichever is higher.

*NEC* Chapter 1, "General," as we have seen, contains fundamental rules that are applicable to all electrical installations. Chapter 2, "Wiring and Protection," continues in this mode, and we shall go through it, extracting the material that is applicable to residential work.

## Color-Coding Details

*NEC* Chapter 2 opens with a discussion of color-coding. In doing any electrical project, it is essential that the correct color be used for each and every conductor.

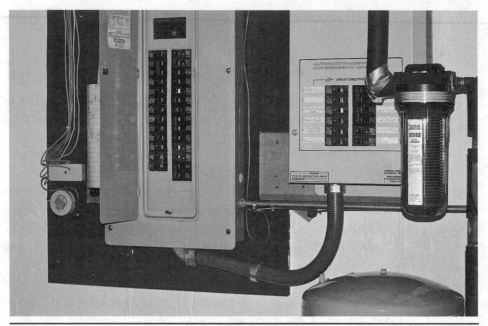

**FIGURE 5-4**    Note the clear violation of working-space requirements.

This is necessary so that the electrician can keep track of what is being done, as well as to facilitate any future maintenance, repair, or additions to the wiring system.

The basic idea behind color-coding—black is hot, white is ground—is very simple, but in actual practice, there are subtle distinctions and exceptions that need to be noted. From the point of view of safety and functionality, the most fundamental distinction is between grounded and ungrounded conductors. Grounding conductors (which are at the same voltage potential as grounded conductors but serve a different purpose) are color-coded differently so that they can be distinguished.

The method for identifying grounded conductors depends on the size. If the conductor is 6 American Wire Gauge (AWG) or smaller, an insulated grounded conductor must be white or gray or have three continuous white stripes in other than green insulation. Some years ago, the *NEC* referred to *natural gray* insulation, but that term was discontinued because it would imply that there was such a thing as "unnatural gray" that could not be used to denote a grounded conductor. Gray may have been used in the past to label ungrounded conductors, so care must be used in working on older systems.

It is not permitted to reidentify an ungrounded conductor as grounded for 6 AWG and smaller wires, except for mineral-insulated metal-sheathed cable (Type

MI), single-conductor photovoltaic outdoor-rated sunlight-resistant wire, and fixture wire. Aerial cable can have a ridge along its entire length to denote that it is grounded.

In sizes larger than 6 AWG, there are different *NEC* rules. The reason is that it would not be practical to stock these larger sizes in all colors. There would be a greater amount of waste, with numerous useless cutoffs instead of one for each job. Conductors larger than 6 AWG may be reidentified as white at terminations only, so it is possible to make use of black wire for any of the conductors. To reidentify, use white paint or, preferably, white phase tape. This must completely encircle the conductor, but it is not necessary to make more than one ring. When grounded conductors pass through an enclosure without termination, reidentification is not required.

For conductors of any size, an insulated grounded conductor in a flexible cord may be identified by a white or gray outer finish, white or gray braid or white or gray tracer in the braid, or a white or gray separator. If in a cord the insulation is integral with the jacket, individual tinned strands will suffice to identify the grounded conductor. Also, ridges, grooves, or white stripes on the outside of a cord will serve to identify a grounded conductor.

If more than one system is included in an individual raceway, cable, or enclosure, each grounded conductor is required to be identified separately, and each of them must conform to the preceding requirements. This coding is to be posted at all branch-circuit panel boards. Although it is not required by the *NEC*, the general trade practice is to use white for the lower-voltage and gray for the higher-voltage grounded conductors.

In switch loops, white wires may be used as ungrounded conductors, but they must be reidentified by marking them black. At one time, the white ungrounded conductor inside the switch enclosure did not have to be reidentified because this was considered self-evident. Current editions of the *NEC* require that both ends of the white conductor in a switch loop be reidentified. In many but not all instances, this has become a moot point because an unused white neutral is required in switch-loop enclosures for future use, and 14-3 Romex is used, but there are exceptions where 12-2 Romex persists, and this is where the reidentification is necessary.

## Bare and Green Wires

Equipment-grounding conductors are identified differently. They originate in the entrance panel, connected to the neutral by means of the main bonding jumper.

From there they accompany the other conductors to each outlet, never again to be electrically connected to the grounded conductors.

Grounding conductors may be bare or insulated. If they are insulated, the color is green. Like grounded conductors, grounding conductors larger than 6 AWG are permitted to be reidentified. This may be accomplished by stripping the insulation from the entire length, in effect converting the conductor to bare, or by coloring the insulation green using paint or, preferred, phase tape. The reidentification must entirely encircle the conductor as opposed to making just a spot. This reidentification is not required in enclosures such as conduit bodies that contain no splices. In flexible cord, a bare conductor qualifies. If it is insulated, the color must be green.

For ungrounded, so-called hot wires, conductors must be clearly distinguishable from grounded and grounding conductors. Accordingly, they may be any color but white, gray, or green and, of course, never bare. When installing wire in raceways, it is possible to pull whatever colors are required. In residential work, Romex is generally used, and you have to work with the colors that are available. For this reason, reidentification may be needed.

Article 210, "Branch Circuits," contains many provisions that pertain to residential installations, and before doing any wiring, the home crafter-electrician should go through this article carefully. Section 210.3, "Rating," states that the ratings for other than individual branch circuits are 15, 20, 30, 40, and 50 amperes. Outside the service, these are the sizes most used in residential work. It is to be emphasized that circuit ratings are determined by the overcurrent device rating, and the conductor size follows from that, as set forth in *NEC* Chapter 3. If, for any reason, for example, to mitigate voltage drop, it is decided to go to a larger wire size, this is permissible because the Code-mandated wire sizes are minimum wire sizes. In such cases, however, the circuit size does not change. It is still determined by the overcurrent device rating.

*NEC* Section 210.4 permits the use of multiwire branch circuits. Some electricians make very extensive use of them because they allow for significant savings in materials and labor. A multiwire branch circuit is characterized by the use of a single grounded (white) conductor as the neutral for two separate circuits with different ungrounded (black and red) conductors. The grounded conductor may be thought of as a shared neutral. All three are contained in a single cable or raceway, in residential work usually 12-3 or 14-3 Romex.

The two hot wires, black and red, are protected against overcurrent at 15 or 20 amperes, and for this reason, each of them, along with the neutral that they share, comprises a separate individual branch circuit. If each of these hot wires were to be loaded close to ampacity, let us say at the full 15-ampere rating of the circuit, would not the neutral be overloaded at 30 amperes so that it could become red hot in a wall cavity and ignite nearby combustible material or at the least undergo insulation damage?

The answer is no. In a properly constructed multiwire branch circuit, the two hot wires are connected to the opposite phase legs in the entrance panel or load center. Because the two neutrals are 180 degrees out of phase, the smaller is subtracted from the larger to find the current on the common neutral conductor. If the two loads are each 15 amperes or any other equal quantities, the current on the single neutral will be 0 amperes. The maximum current in the common neutral occurs when the load on one circuit is maximum, 15 amperes in this example, and the load on the other is 0 amperes. In this situation, the current in the shared neutral is 15 amperes, which is not a hazardous overload.

Several savings result from the use of a multiwire branch circuit. First is the obvious savings in wire, because a single white conductor serves as the neutral for two circuits. Also, there is a savings in certain hardware because a single cable connector, run of staples, and so on is needed.

In a raceway system, there is less conduit fill, which may translate into a smaller raceway size. There is less labor in the conduit pull. In a cable system, there is less labor because one cable run replaces two, with less drilling of studs and fewer terminations to be made. Moreover, the round three-wire cable is easier to run than the flat two-wire cable because you don't expend so much effort preventing cable twists. For these reasons, some of the high-volume electrical contractors favor the use of multiwire branch circuits where they are permitted. A few dollars saved on every job adds up to a fortune in the life of a company.

There is a major downside to multiwire branch circuits, and this is why many of the better electricians will not use them. If, in the future, in the course of repair or alterations, a less knowledgeable worker shifts one of the hot legs to the wrong phase in a panel, the currents in the neutral suddenly will become additive, and there is the possibility that loaded beyond its overcurrent protection, that neutral will overhead and ignite nearby combustible material.

The *NEC* recognizes the potential for this hazard and rules are in place to mitigate it:

> Each multi-wire branch circuit is to be provided with a means that will simultaneously disconnect all ungrounded conductors at the point where the branch circuit originates. This is usually a double-pole breaker, as normally used for a 240-volt circuit. Two single-pole breakers with a listed handle tie to ensure that they trip simultaneously will serve the purpose, but the usual procedure is to use a double-

pole breaker. Under no circumstances is an improvised expedient such as a nail or piece of wire through the holes in the single-pole breakers to be substituted.

Another *NEC* safeguard for multiwire branch circuits is the requirement in Section 210.4(D), "Grouping," which states that the ungrounded and grounded circuit conductors of each multiwire branch circuit are to be grouped by cable ties or similar means in at least one location within the panel board or other point of origination.

These rules are designed to ensure that the two hot legs of a multiwire branch circuit will not be connected to the same phase in the future. But a little knowledge is dangerous, and in the fullness of time, it is a distinct possibility that a partially trained worker will swing one of the phase wires over to the wrong leg. So it is not recommended that the home crafter-electrician use multiwire branch circuits in the initial installation.

## Ampacities and Device Ratings

*NEC* Chapter 2 moves on to lay out some additional requirements that apply to residential electrical installations. Branch circuits, unless otherwise specified, are to have an ampacity sufficient for the loads served and are not to be smaller than 14 AWG. Where a branch circuit supplies continuous loads or any combination of continuous and noncontinuous loads, the rating of the overcurrent device is to be not less than the noncontinuous load plus 125 percent of the continuous load. Table 210.21(B)(3), "Receptacle Ratings for Various Size Circuits," presents information that is somewhat counterintuitive, as shown in Table 5-2.

We ordinarily believe that the overcurrent device should be the weakest link in the chain so that it will shut down the circuit before any other parts of it are stressed. As a rule, this is true, but there are many exceptions throughout the

**TABLE 5-2   Circuit Ratings for Various Size Circuits**

| Circuit Rating (Amperes) | Receptacle Rating (Amperes) |
| --- | --- |
| 15 | Not over 15 |
| 20 | 15 or 20 |
| 30 | 30 |
| 40 | 40 or 50 |
| 50 | 50 |

Code, and among them are receptacle ratings. One of the most important of them is the fact that a 15-ampere receptacle is permitted on a 20-ampere circuit. As mentioned previously, many electricians use 12 AWG copper Romex for almost all residential circuits, even when the loading would allow for 14 AWG copper Romex. Then there is the option to use 15- or 20-ampere breakers for overcurrent protection. The 15-ampere breaker is more sensitive, offering more robust overcurrent protection, whereas the 20-ampere breaker is more resistant to nuisance tripping, permitting the full benefit of the 12 AWG copper Romex, which has higher ampacity. Table 210.21(B)(3) permits 15-ampere receptacles, shown in Figure 5-5, to be used with either size circuit.

Chapter 2, Part III, "Required Outlets," goes into considerable detail concerning required receptacle placement and spacing within a residence. In wiring habitable rooms, laundries, bathrooms, and kitchens, it is necessary to have this part of *NEC* Chapter 2 before you and to use it as a checklist to make sure that everything is in place.

**Figure 5-5** Receptacles are available in 15- and 20-ampere ratings as well as other sizes.

*NEC* Chapter 3, "Wiring Methods and Materials," continues the coverage of general principles underlying all electrical work, as seen in Chapters 1 and 2. A lot of this material pertains to residential wiring, and we shall summarize the topics that are of interest to the home crafter-electrician

Section 300.5, "Underground Installations," contains information needed to build an underground service. Also, the time may come when you will want to bury an electrical feeder that supplies power to an outbuilding, outdoor light fixture, septic pump, or outdoor power outlet that is not attached to a building. You will need to refer to Table 300.5, "Minimum Cover Requirements." All parts of this table except for the bottom row are relevant to residential electrical work. Notice that there is not an entry specific to services. However, utilities specify burial depths and other requirements.

Notice that where conductors are placed in RMC, minimum cover is reduced to as little as 6 inches (or even 4 inches with concrete), so this is helpful where bedrock is encountered. Some types of wires, either individual conductors or in cable, are rated for direct burial without raceways. An example is underground feeder (Type UF). The preferred installation, however, is in conduit, usually polyvinyl chloride (PVC). This arrangement has the advantage that if conductor replacement becomes necessary, it can be done without digging, simply by pulling in new wire.

If the conductors are buried directly, they still must be protected by raceways where they emerge from grade either to where they enter the building or to a height of 8 feet above grade. The best way to bring buried conductors into a building is to have them emerge from underground and enter the building through a Type LB fitting, as shown in Figure 5-6, as opposed to going through a hole in the concrete.

The interiors of enclosures or raceways installed underground are considered to be wet locations, so suitable conductors and splicing methods are required. If moisture can enter and contact live electrical parts, the raceway is to be sealed at either or both ends. In addition, raceways are to be provided with expansion joints if needed to compensate for thermal expansion and contraction.

Article 310, "Conductors for General Wiring," is perhaps the most referenced part of the entire *NEC*. With its many tables pertaining to conductor sizing, it is not practical to memorize individual requirements, nor is this necessary because the Code can be used on an open-book basis.

The heart of Article 310 is a series of tables, 310.15(B)(16) through 310.15(B)(21). They are used to size conductors in most electrical installations, and only by understanding and applying the tables with accompanying notes can we be certain that the wires have sufficient ampacity to carry the current without overheating.

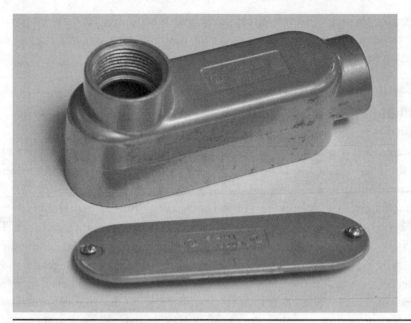

**FIGURE 5-6** Type LB is one of several types of conduit fittings that facilitate conductor pulls and make angle turns.

Each of these tables has a lengthy heading that details the conditions for which the table is applicable. By far the most used is Table 310.15(B)(16). This table gives allowable ampacities, the maximum current that can be carried safely without danger of overheating. The table applies to insulated conductors rated up to and including 2,000 volts, rated 60 through 90°C, where there are not more than three current-carrying conductors in raceway or cable or buried directly in earth, based on an ambient temperature of 30°C.

Other tables in this series give ampacities for over 2,000 volts and for free conductors in air, that is, aerial installations. It is the first table, though, that is applicable for most residential jobs. Most of the time, you will use this table in reverse, beginning with the ampacity that will be needed, then working back to the top row to choose the type of conductor that is to be used, and then proceeding to the column at the far left. It will tell you, above all, what you need to know, which is the correct wire size.

The process is quite simple once you have chosen the correct table based on the criteria in the heading. The left half of the table contains ampacities for copper conductors, and on the right are aluminum conductor ampacities, rarely used in residential work except for services. Also, for residential work, you will be referring to the top half of the table for conductor sizes not over 4/0 AWG. This is all there is to choosing wire size, except that before beginning, there are operations to be performed in the course of finding the desired ampacity, which is the maximum current the conductor will be required to carry.

## Correction and Adjustment Factors

First, determine the load. Obtain the total based on the general lighting load from *NEC* Table 220.12, as explained earlier, and the nameplate rating or marking on any equipment that will comprise the load. To this amount, various derating factors may be applied, as explained in our coverage of *NEC* Chapter 2.

The quantities so determined are matched up to the ampacities in Table 310.15(B)(16) or other applicable table so as to find the required conductor size for the wire type that is to be used. Before the final decision can be made, though, it is necessary to alter the ampacity to suit certain conditions pertaining to the installation. This involves, where necessary, multiplying the ampacity successively by one or more numbers that are given in *NEC* tables. These numbers are less than 1 (i.e., percentages) that cause the ampacity to become lower, meaning that you may have to choose a larger conductor size to provide the minimum required ampacity. Because the factors are multipliers, the order in which they are applied does not matter. We will discuss them in the order in which they appear in the *NEC*. See Table 5-3.

Table 310.15(B)(2)(a), "Ambient Temperature Correction Factors Based on 30°C," provides multipliers that are used to reduce the ampacity so that in some cases larger conductors are needed. The ampacities given in Table 310.15(B)(16) are to be multiplied by the factors shown in this table before choosing the conductor size. Because this particular ampacity table, shown in Table 5-3, is based on an ambient temperature of 30°C, we use these correction factors rather then those in Table 310.15(B)(2)(b), which is based on 40°C. In other words, be certain that both tables are based on the same ambient temperature or the results will not be valid.

Temperature correction factors must be applied when conductors, not necessarily the load, are exposed to high-temperature surroundings somewhere along the line, for example, when a branch circuit is routed near a heat source such as a furnace or uninsulated heat duct. The high ambient temperature will interfere with dissipation of heat due to current carried by the conductor, making for an increase in temperature. The result, if conductor size is not increased, could be fire hazard and/or damage and premature failure of the conductor insulation. By applying the correction factor, which may dictate a larger conductor size, the hazard is mitigated.

**TABLE 5-3** Ambient Temperature Correction Factors

| Ambient Temperature (°C) | Conductor Temperature 60°C | Conductor Temperature 75°C | Conductor Temperature 90°C |
|---|---|---|---|
| 10 or less | 1.29 | 1.20 | 1.15 |
| 11–15 | 1.22 | 1.15 | 1.12 |
| 16–20 | 1.15 | 1.11 | 1.08 |
| 21–25 | 1.08 | 1.05 | 1.04 |
| 26–30 | 1.00 | 1.00 | 1.00 |
| 31–35 | 0.91 | 0.94 | 0.96 |
| 36–40 | 0.82 | 0.88 | 0.91 |
| 41–45 | 0.71 | 0.82 | 0.87 |
| 46–50 | 0.58 | 0.75 | 0.82 |
| 51–55 | 0.41 | 0.67 | 0.76 |
| 56–60 | — | 0.58 | 0.71 |
| 61–65 | — | 0.47 | 0.65 |
| 66–70 | — | 0.33 | 0.58 |
| 71–75 | — | — | 0.50 |
| 76–80 | — | — | 0.41 |
| 81–85 | — | — | 0.29 |

Table 310.15(B)(3)(a), "Adjustment Factors for More than Three Current-Carrying Conductors," shown in Table 5-4, gives another factor that sometimes must be applied to the ampacity so that conductors can be chosen that will not experience heat rise.

The accompanying notes provide interpretations on how to apply the table. The question that arises is how to determine whether or not a conductor is to be considered current-carrying.

**TABLE 5-4** Adjustment Factors When More than Three Current-Carrying Conductors Are Used

| Number of Conductors | Adjustment Factor (%) |
|---|---|
| 4–6 | 80 |
| 7–9 | 70 |
| 10–20 | 50 |
| 21–30 | 45 |
| 31–40 | 40 |
| 41 and above | 35 |

Equipment-grounding and bonding conductors are not counted as current-carrying conductors. Neutral conductors that carry only the unbalanced current from other conductors of the same circuit are not required to be counted as current-carrying conductors. Conductors that are connected to electrical components but cannot be simultaneously energized are not required to be counted. Spare conductors must be counted. Adjustment factors are not required to be applied to conductors in raceways having a length not exceeding 24 inches.

The temperature factors are known as *correction factors*. For conductor count, the factors are known as *adjustment factors*. This difference in semantics helps to distinguish the two types of multipliers. They are both derived from their respective tables and are applied in the same way to the ampacities in the Chapter 3 ampacity tables.

## Box-Fill Calculations

Article 314, "Outlet, Device, Pull, and Junction Boxes; Conduit Bodies; Fittings; and Hand Holes," contains box-fill rules. The basis for these rules is the statement that boxes and conduit bodies are to be of a size to provide free space for all enclosed conductors. In no case is the volume of the box as calculated in Section 314.16(A) to be less than the fill as calculated in Section 314.16(B).

Boxes and conduit bodies enclosing 4 AWG or larger conductors are figured differently and must comply with Section 314.28. These sizes are rarely seen in residential work except in services, which are not spliced.

Boxes that are overfilled create a severe hazard, and this is a source of difficulty in many faulty installations. Where physical force must be used to get all the conductors into a junction box so that the cover can be attached or to pack the conductors into a wall box so that a switch or receptacle, particularly a GFCI (because it is larger), can be bolted into place, there is the possibility that the conductors will be damaged or the termination will be loosened or pulled apart. Then there is a chance of an arcing fault, or the device or load could stop working in the future. Wire nuts, shown in Figure 5-7, especially if they are used to join more than two conductors, are prone to eject conductors when forced into an overcrowded box.

Electricians quickly get a sense of what will go into a box, but in borderline cases or unusually large boxes, it is possible to make a mistake. Use of deep wall boxes and 4 × 4 square boxes, as shown in Figure 5-8, rather than octagonal junction boxes is helpful. The problem is that when it is time to terminate the device and insert it into or on the box along with associated wiring and devices, the finish

**FIGURE 5-7**   Wire nuts contribute to box fill and must be factored into the equation.

**FIGURE 5-8**   Large enclosures make for less box fill so that the installation is compliant and there is no chance of conductor damage.

wall or ceiling material is already in place, and it is not possible to replace the box with a larger one without tearing out the finish material.

The remedy for this sort of problem is to perform the box volume and fill calculations so as to determine prior to installation whether a proposed box size is suitable for the intended use. Section 314.16(A), "Box Volume Calculations," states that the volume of a wiring enclosure (box) is to be the total volume of the assembled sections and, where used, the space provided by plaster rings, domed covers, extension rings, and so forth, which are marked with their volume or are made for boxes the dimensions of which are listed in Table 314.16(A), "Metal Boxes."

Section 314.15(B), "Box-Fill Calculations," contains five paragraphs that describe how to calculate the box fill of various types of hardware. Some small fittings, such as locknuts and bushings, require no fill allowance.

Table 314.16(B), "Volume Allowance Required per Conductor," lists an amount for each conductor. All you have to do is make a list of what is to go into the box and determine the size of the box needed.

The rest of Chapter 3 is made up of Articles 320 through 399, which cover specific varieties of cable and raceways. The articles follow a common organizational template, making them easy to navigate. Articles 320 through 340 cover cable types in alphabetical order. These are followed by raceways, which are subdivided into types of tubing and conduit, but they are not in alphabetical order. Finally, there are four additional kinds of wiring.

The best way to find the article that covers a desired type of cable or raceway is to look it up in the Table of Contents at the beginning of the *NEC*. Some of the cable and raceway types are used very frequently, and you will quickly learn the article numbers that pertain to them, for example, Type NM (Romex), which is covered in Article 334.

*NEC* Chapter 4, "Equipment for General Use," contains articles on specific devices and equipment such as switches, receptacles, light fixtures, appliances, fixed space heating, motors, air-conditioning equipment, and so forth. A lot of this material has to do with construction specifications and is of interest primarily to manufacturers. Some sections, such as Article 430, "Motors, Motor Circuits, and Controllers," pertain to industrial facilities, and there is little relevance to residential work. The home crafter-electrician, however, will want to take a good hard look at some parts of Chapter 4, for example, Section 410.16, "Luminaires in Clothes Closets" (*luminaire* is the Code term for a light fixture).

A light fixture inside a clothes closet is a potential fire hazard because the stored clothing is highly flammable, and the luminaire is a source of heat, both in normal operation and in the case of a malfunction where sparks may be emitted.

Section 410.16 contains a detailed list of clearances that must be observed during the initial installation. In some closets, because of limited size, it is impossible to comply with these clearances. In such cases, there is the option of installing a battery-powered closet light, made for the purpose, that is incapable of igniting nearby combustible materials. Another work-around is to position a ceiling light just outside the closet in the adjoining room.

Section 410.16 contains a detailed diagram showing all clearances. The basic clearances between luminaires and the nearest point of closet storage space are

- Twelve inches for surface-mounted incandescent or light-emitting-diode (LED) luminaires with a completely enclosed light source installed on the wall above the door or on the ceiling
- Six inches for surface-mounted fluorescent luminaires installed on the wall above the door or on the ceiling
- Six inches for recessed incandescent or LED luminaires with a completely enclosed light source installed in the wall or ceiling
- Six inches for recessed fluorescent luminaires installed in the wall or ceiling
- Surface-mounted fluorescent or LED luminaires are permitted to be installed within the closet storage space where identified for this use.

*NEC* Chapter 4 goes on to cover some other types of equipment for general use. Still on the topic of luminaires, there are parts of the article that are concerned with supports. A luminaire that weighs over 6 pounds or exceeds 16 inches in any dimension is not to be supported by the screw shell of a lamp holder. Another issue regarding the installation of luminaires is grounding. All exposed normally nonconducting metal parts are to be mechanically attached to the equipment-grounding conductor that accompanies the branch circuit.

After the wiring has been roughed in and wall and ceiling materials are in place, we are ready to start the electrical finish work. A good part involves hanging the wall and ceiling light fixtures. If they are new, there will be a complete set of installation instructions, so if there is any question concerning electrical connections or mechanical mounting, there will be an answer. Most of this is self-evident. If the light fixture is a heavy chandelier, a helper will be needed to support it while you wire nut the electrical connections prior to mounting it to the ceiling.

It is stated in this article that wiring on or within luminaires is to be neatly arranged and not exposed to physical damage. Excess wiring is to be avoided. Conductors are to be arranged so that they are not exposed to temperatures above those for which they are rated.

FIGURE 5-9   A paddle-fan luminaire requires special mounting arrangements to make sure that it is secure. Consult the manufacturer's documentation before installing the ceiling box.

Many incandescent luminaires come with labels stating a maximum size bulb to be used, often 60 watts. This warning should be observed because the heat from a larger bulb can char nearby combustible material, progressively lowering the ignition temperature.

Section 410.50, "Polarization of Luminaires," such as the combination paddle-fan light fixture shown in Figure 5-9, states that the grounded conductor, where connected to a lampholder, is to be connected to the screw shell.

The installation, when completed, should consist of the equipment-grounding conductor connected to the outer housing and the grounded conductor connected to the screw shell. These two parts are to be insulated from each other so that there is no electrical connection between grounded conductor and grounding conductor downstream from the entrance panel, where they are joined by the main bonding jumper (unless there is a main disconnect in a separate enclosure still farther upstream). Such a connection would violate one of the most basic principles of all grounded wiring systems. Under no circumstances is the hot (black) conductor to be connected to the screw shell or to the fixture housing.

Article 422, "Appliances," covers this type of equipment, defined in Article 100, "Definitions," as utilization equipment, generally other than industrial, that is normally built in standardized sizes or types and is installed or connected as a unit to perform one or more functions, such as clothes washing, air conditioning,

**FIGURE 5-10** *NEC* Article 422 covers appliances such as this home refrigerator.

food mixing, deep frying, and so forth. A typical home appliance is shown is Figure 5-10.

An appliance may be portable or fixed, and it may be hardwired or cord-and-plug connected. Some appliances, such as an electric range with oven or a tankless hot-water heater, shown in Figure 5-11, consume significant amounts of power and are among the largest loads in a home. Others, such as an electric refrigerator, consume modest amounts of power and may be connected to an ordinary 15-ampere branch circuit.

A key provision of this article is that the branch-circuit rating for any appliance other than a motor-operated appliance that is a continuous load is to be 125 percent of the marked rating. *Continuous* is defined as expected to operate for 3 hours. A fixed storage-type water heater that has a capacity of 120 gallons or less is to be considered a continuous load for the purpose of sizing branch circuits.

**FIGURE 5-11**    A tankless (quick-recovery) hot-water heater draws more current but for shorter durations than a conventional hot-water heater.

## Appliance Identification

To wire an appliance, you need to find the rating of the unit. This will appear on the nameplate. The nameplate is a small metal label that is permanently attached to the appliance. It is usually on the back, inside a door, or at some other location that is not difficult to find. Always begin with the nameplate.

The nameplate contains certain information that is prescribed by the *NEC*. This information differs depending on the type of equipment. Motor nameplates always state the revolutions per minute (rpms), but this is obviously not relevant for many appliances. The nameplate of a typical quartz infrared portable heater states the name of the manufacturer; the voltage, frequency, and power rating in watts; the model number and date of manufacture; and a logo showing that it is Underwriters Laboratories (UL) rated.

An important set of provisions that pertains to all appliances is contained in *NEC* Part III, "Disconnecting Means." Section 422.30, "General," states that a means is to be provided to simultaneously disconnect each appliance from all ungrounded conductors. If an appliance is supplied by more than one branch circuit or feeder, these disconnecting means must be grouped and identified as the appliance disconnect.

Section 422.31, "Disconnection of Permanently Connected Appliances," states that for permanently connected appliances rated at not over 300 volt-amperes or $\frac{1}{8}$ horsepower, the branch-circuit overcurrent device is permitted to serve as the disconnecting means. For permanently connected appliances rated over 300 volt-amperes, the branch-circuit switch or circuit breaker is permitted to serve as the disconnecting means where the switch or circuit breaker is within sight of the appliance or is lockable.

Section 422.33, "Disconnection of Cord and Plug-Connected Appliances," states that an accessible separable connector or an accessible plug and receptacle are permitted to serve as the disconnecting means. A unit switch, such as on the front panel of an arc welder in a shop, associated with a single-family home is permitted to be the disconnecting means provided that there is another disconnecting means, and the other disconnecting means does not have to be within sight of the appliance.

Article 424, "Fixed Electric Space-Heating Equipment," includes coverage of the very common electric baseboard heat units, as shown in Figure 5-12. Often the heat source, such as an oil furnace, is adequate for an existing building, but when an addition is built, the furnace may not have the capacity to heat the new space. Rather than upgrading to a larger furnace or possibly building a new chimney and installing a second furnace, the inexpensive option is to put baseboard electric heat in the new addition. A single zone will do for a one-room addition.

Electric baseboard heat is available in various sizes, the electrical rating corresponding to the lineal length. These units may be obtained in low- and high-density models. The low-density electric baseboard heat equipment operates at a lower temperature, but the tradeoff is that for a given wattage, there is greater length.

If electric baseboard heat is to be installed, it is necessary to start with a design. Some utilities provide assistance. An engineer will visit the site and perform a heat-loss survey. This involves taking into consideration the area in square feet of the addition or complete building; its shape; the thickness of the walls, ceilings, and floors and the amount of insulation; the type of foundation; the number, type, and total area of windows; and above all, the climate, with the lowest expected ambient temperature.

**FIGURE 5-12**    An electric baseboard space heater. The receptacle placed above it is a common listing violation because the cord is likely to enter the enclosure and be damaged by the heat.

The heat-loss survey will provide the information needed to determine the total amount of electric heat needed in watts. From this figure and the wall lineage, not including doors and any floor-to-ceiling glass, it is possible to determine size and placement of the strips. Electric baseboard heating units normally are centered under windows so that the rising heat will minimize moisture and ice.

An important consideration in planning an electric baseboard heat installation is that receptacles are not to be installed above the units, a very common violation. The reason for this is that power cords and extension cords plugged into a receptacle that is located above a baseboard strip could find their way into a heating enclosure and be damaged by the heat. Because receptacles must be placed at prescribed intervals, there is the potential for conflict. So receptacle and heat-strip placements have to be coordinated. For a problem installation, baseboard heating units are available that incorporate a built-in receptacle at the midpoint. When wiring them, connect the receptacles to the appropriate branch circuits, as opposed to bugging them off individual legs of the 240-volt heating circuits.

A baseboard heating unit includes the electrical wiring enclosure, so in roughing in the wiring, it is not necessary to install a wall box. Just leave a 10-inch whip emerging from a hole in the wall material. This hole should be located at the end

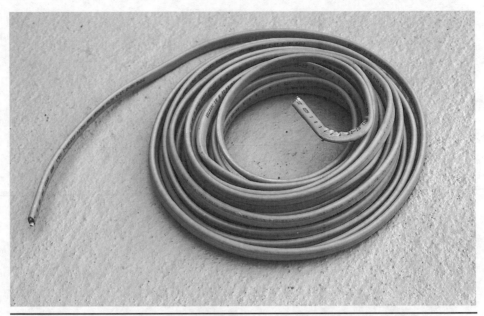

**Figure 5-13**   Romex cable is used extensively in residential wiring.

of the electric baseboard strip, lining up with the wiring compartment. The hole will be concealed behind the unit, so your electrical work will be invisible.

Romex 12-2, shown in Figure 5-13, is used in a residence for most baseboard heating applications. Units can be daisy-chained to take advantage of the full ampacity, but don't forget that this type of load is considered continuous, so it has to be multiplied by 1.25.

The total heat load is divided into zones. If the addition is a small room, there will be a single zone. A full house may have any number of zones, limited only by the capacity of the entrance panel.

Each zone has its own thermostat. The simplest type of thermostat circuit is in-line—two wires from the entrance panel or load center in and two wires out to the first baseboard heater. Thermostat input terminals are marked "line," and output terminals are marked "load." There is no need to match up the legs, and of course, there is no neutral.

The two conductors are connected at the breaker box to a 20-ampere double-pole overcurrent device. If Romex is used, the white must be reidentified using any color except white or green. This must be done at the breaker, thermostat, and baseboard heater.

An in-line thermostat has the disadvantage that the entire current drawn by the load must pass through the thermostat, and it must be switched by it. This means that the in-line thermostat will have a finite life, and the end-of-life event

**FIGURE 5-14**   A thermostat should be installed at eye level. A standard wall box can be used.

may be a terminal failure, accompanied by heat. For this reason, it is important to provide a metal wall box for the thermostat. The alternative is to have the in-line thermostat be a low-voltage device, as shown in Figure 5-14.

The thermostat should be suitably located so that the temperature will be representative of the zone. The thermostat should not be on a wall where direct sunlight will cause it to turn off the heat, and it should not be in a drafty location by an exterior door. For the low-voltage control circuit, a transformer is required, as shown in Figure 5-15.

## *NEC* Article 430, "Motors, Motor Circuits, and Controllers"

Everyone has noticed that once in a while, when a heavy load is switched on, the lights in the house dim for a second or so. If this is excessive, taking place when even modest loads are turned on, it can be a sign that the service or branch circuit is undersized. It is a consequence of Ohm's law. When the amperes increase and there is resistance in a circuit, the voltage measured downstream from the resistance drops. This causes the lights to dim.

It is also a consequence of the fact that most loads draw more current when first powered up. This is true of simple resistive loads such as incandescent light-

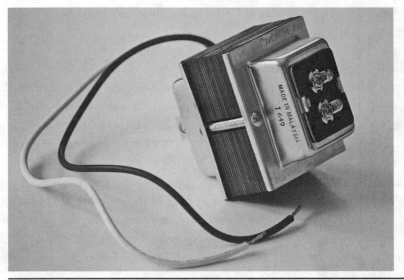

**FIGURE 5-15**   The control-circuit transformer is located at the furnace.

ing and heating elements, but it is even more true of motors. Due to inertia of rest, it takes a while for the rotor to get up to speed, and during this time, the windings produce heat rather than the full-rated revolutions per minute. The same thing happens when the motor's shaft is, for any reason, locked up and unable to turn. A motor with a locked rotor will quickly overheat unless it incorporates high-impedance windings. Step motors are like this, capable of high holding torque while not turning.

In the home, most motors are cord-and-plug connected. For the home crafter-electrician, wiring most motors is no more intricate than plugging them into a receptacle that is on a 15- or 20-ampere branch circuit. But there may come a time when you need to wire a large motor in an attached shop or garage, let's say for an air compressor or a 12-inch wood planer. There will be a bit of a learning curve. If you proceed with the installation as you would for a comparable nonmotor load, there is a high probability that the overcurrent device will trip out within 2 or 3 seconds before the motor gets up to speed.

## Motor Fundamentals

*NEC* Article 430, "Motors, Motor Circuits, and Controllers," provides guidance. If you understand the rules for various sizes and types of motors, and if you apply them accurately, the installation will perform well. (That is, unless the motor is an old junker with worn-out bearings or winding insulation that permits excessive current leakage to ground.)

By way of background, we'll review the primary types of motors, along with the basic principles that are common to all of them and make them work. A *motor* is a device that converts electrical energy into useful mechanical motion. By this definition, a loud speaker is a motor, but we are usually thinking of a rotary machine with one or two output shafts. These may be attached to pulleys, gears, saw blades, grinding wheels, hydraulic pumps, or many other types of devices that perform work for us.

Rotary electric motors consist of stators and rotors. The stator windings are mounted to the inside of the case. At both ends of this housing are bearings. The outer part of each bearing is held firmly in place so that it cannot slip or vibrate, and the inner parts of the two bearings, along with the shafts, are free to turn. One or both shafts extend through the bearing(s) and emerge through the end housings into the outer world.

The magnetic interaction of rotor and stator is what makes the shaft turn. The polarity of the magnetic field of either the rotor or the stator has to be reversed continually in order for the motor to turn. One but not both of these magnetic fields may be created by a permanent magnet. From time to time, a scheme will surface whereby both rotor and stator consist of permanent magnets, and perpetual motion is achieved with no external electrical supply required. This can't happen. The polarity of either the stator or the rotor must be reversed periodically for rotation to take place. The only way to obtain this reversal would be for one set of permanent magnets to be physically flipped over, and this process would require whatever energy could be extracted from the output shaft. At best, the motor would rotate one-half turn and then stop, never to move again.

The spinning of the shaft of a rotary electric motor is made possible by a continuous input of electrical energy. Either the rotor or the stator must consist of electromagnetic windings that create a magnetic flux in a core, usually soft iron, that is magnetically permeable. A magnetic field is established with regularly reversing polarity because of the pulsating electricity supplied to the windings.

Consequently, the magnetic field of the rotor is continuously chasing after the magnetic field of the stator, causing the rotor to turn. To make the magnetic field reverse polarity, the electric current must reverse polarity, and this happens by a process known as *commutation*. The commutation can be internal (inside the housing of the motor, made possible by a *commutator*) or outside the motor, in which case the pulsations are the work of the utility-supplied alternating-current (ac) line voltage or are created by a local electronic oscillator or mechanical vibrator that is associated with the motor but outside the unit.

There are several types of motors, and what makes them differ is the commutation strategy and how it plays out in the rotor-stator interaction. The simplest type of rotary motor is the internally commutated direct-current (dc) motor. This was the first type of motor to be developed in part because prior to the Tesla-Westinghouse implementation of alternating current, only dc was available. The source was very inefficient short-lived chemical batteries, and for this reason, the first dc motors were not very powerful. Because they could not do much useful work, they were good only as amusing toys and scientific curiosities. With Thomas Edison's early dynamo-powered electrical distribution system, dc motors suddenly gained great importance in the industrial world. With the arrival on the scene of the ac induction motor, the dc motor has lost ground but still occupies an important place in niche applications.

## What Makes a DC Motor Turn?

In a dc motor, a nonpulsating voltage is applied to the terminals of the motor. The dc is applied to the stator windings, and a stationary magnetic field is established. The dc is also applied to the rotor windings. For this type of motor, rotor and stator windings are wired in parallel, and a variable resistance is put in series with the field windings to control the speed. The whole thing is easy to check out with an ohmmeter. Expect to see low resistance even with the variable resistance at its highest setting. If you can disconnect one pole of the variable resistance, each of the components can be measured individually.

Voltage has to be applied to the spinning rotor. If the wires are connected directly, they very quickly twist and break-off. This is the dilemma with any motor—how to get electrical energy into the rotor or otherwise make it magnetically active. In the dc motor's most basic form, the input for the armature is applied by direct electrical connection to the brushes. These were originally actual copper brushes positioned so as to contact the different segments of the commutator, causing the current to flow in the correct sequence and polarity to the rotor windings.

Metal brushes have been replaced by carbon brushes, which have the advantage that they exhibit less sparking. They point straight toward the center of the commutator. The end of the brush is manufactured with a curvature that is compatible with the commutator. At the other end of the brush, inside the brush holder, is a light spring that maintains just the right pressure of the brush on the spinning commutator, advancing the brush a slight amount to compensate for wear.

These carbon brushes wear down eventually and must be replaced. If the motor has been running and has not overheated but refuses to start, the first thing to look for after a bad electrical connection has been ruled out is the brushes. If one or both appear to be too short, broken, pitted, or crumbly, both should be replaced. This usually can be done quite easily with no disassembly of the motor.

Exposure to moisture or oil from outside contamination or excessive bearing lubrication will soften the brushes and cause premature wear. If the brushes look good, sometimes the springs have lost their tension, and operation can be restored by stretching them.

Replacement brushes are readily available for most motors. Some technicians have made a hard-to-find brush by cutting down a larger one, but don't forget, the curvature has to match that of the commutator. Also, small amounts of metal are added to the brush material, and it is heat treated in various ways during manufacture, so the electrical characteristics may not be suitable for the motor.

If the motor is not an enclosed type, it may be possible to see the brushes in action while the motor is running. A slight amount of sparking is normal, but when it becomes excessive, the brushes are worn and should be replaced. It is false economy to run the brushes to failure because commutator damage will result. As sparking increases, there is more heat, and this spells trouble. Before long, the sparking will resemble a bright gas flame, and the commutator will need to be rebuilt or replaced. This involves complete motor disassembly unless it is possible to clean the commutator in place.

If the commutator is badly pitted, the area where the brushes make contact is worn excessively, or if the commutator is out of round, as measured by a micrometer, it will need to be taken to a machine shop or motor rebuilder to be turned on a lathe. Afterward, the insulating material in the gaps between the commutator segments is cut back a slight amount using a tool that resembles a hacksaw blade with a handle. If this is not done, the motor will "eat" brushes. Worn brushes cause commutator wear, and a worn commutator causes premature brush failure. Changing brushes is a simple routine task, whereas rebuilding a commutator is a more extensive project.

Motors that have brushes require periodic inspection and maintenance. Some motors, numerically the vast majority, do not have brushes, and they require much less maintenance.

## An Answer

The brushless dc motor solves this problem by having a rotor equipped with permanent magnets rather than electromagnetic windings. With no electrical connection to the rotor, there are no brushes. Commutation is external. A pulsed voltage of alternating polarities is applied to the field windings. It is created in a local electronic module that is associated with the motor but outside it. Because there are no moving parts except for the spinning rotor, maintenance is minimal.

Dc motors are plentiful in the home. Battery-operated children's toys, computer disk drives, DVDs, inkjet printers, and automobiles all have dc motors. Some can be rebuilt; others have housings that are not intended to be opened. This shouldn't stop you if you have an inquiring mind.

## Steppers and Servos

Stepper motors and servomotors are interesting variants of the basic dc motor described earlier. Both are able to rotate a part turn and then stop with or without holding torque. They are both externally commutated. But the stepper motor is open loop, whereas the servo is closed loop, making for a higher-performance machine.

If, for any reason, the stepper motor falls out of synch, it must be reset. For this reason, it is not suitable in demanding applications. The servo system, in contrast, is equipped with an optical or other sensor attached to the motor or load so that there is continuous feedback. If the motor loses synchronization, there is ongoing error correction.

The stepper motor is externally commutated. Electrical pulses from outside go to the field windings, and the armature, as in other types of motors, chases after the rotating magnetic field that is created. A simple type of stepper is the variable-reluctance motor. It has a soft-iron notched rim that is attached to the outside of the armature. The notched teeth rotate with the armature close to the stationary field windings, separated by a small air gap. These teeth are alternately closer and farther from the stationary field windings. A freely moving body of high magnetic permeability such as soft iron always wants to move so as to create a magnetic flux path with the least reluctance. Reluctance in a magnetic circuit corresponds to resistance in an electrical circuit. Where there is magnetic flux, there is always a circuit. Sometimes, as in a stationary permanent magnet, the circuit is open.

Soft iron is quite permeable to magnetic flux, so there is less reluctance when the protruding parts of the armature pass through the flux lines coming from the stator. The armature will continue to position itself for minimum reluctance in the magnetic circuit. The magnetic field has to turn if step-motor rotation is to occur. For this to take place, the field coils must be energized by varying electrical pulses. An external controller is needed to generate these pulses. There should be two wires for each coil, but usually one side of each coil is connected inside the motor to a common wire.

Step motors have many wires, whereas standard brushed dc motors have two wires. A single pulse will make the motor turn a part turn. Then it will remain stationary with or without holding torque, which can be achieved by dc from the controller. The motor can endure this locked-rotor operation because it has more impedance than other similar-sized motors. Successive part rotations with pauses and reversals can occur in rapid succession as specified by controller programming. Ink-jet printers work in this way, and such behavior is useful in student robotics projects.

A waveform from the controller will cause the step motor to turn at the desired speed, reversing or making fractional turns. It would be possible, using a microphone and appropriate circuitry, to play notes on a musical instrument, causing the step motor to turn at different speeds.

Step motors may be manufactured with various numbers of field coils and notched teeth around the perimeter of the armature. These numbers work together to determine the smallest possible step, often expressed in degrees. For example, a 3.6-degree stepper has a maximum of 100 possible steps. This is also known as the *resolution* of the step motor.

Besides the variable-reluctance model, there are three other types of step motors—permanent magnet, hybrid, and Lavet. Permanent-magnet step motors do not have notched teeth. In their place are rotors with permanent magnets attached. They are characterized by greater dynamic and holding torque, so they would be the logical choice if a heavy load is to be moved or held in place. Hybrid steppers are small and powerful. This combination is made possible by the fact that they combine variable-reluctance and permanent-magnet constructions. The Lavet-type stepping motor turns only in one direction because it is single phase. This type of motor is suitable for wristwatches with quartz clocks because the motor requires little power, making for long battery life. Lavet steppers are also used for automotive dashboard instrumentation. The more robust servomotor, shown in Figure 5-16, is suitable for more demanding applications, such as automotive assembly-line machines connected to programmable logic controllers (PLCs).

Step motors and servos often can be used interchangeably. In small, low-powered applications or where cost is the primary issue, step motors prevail. The servomotor moves in where reliability and robust performance are important and where the higher cost is not prohibitive.

The step motor works well until, due to load binding, electrical fault, or other problems, synchronization is lost. Once this happens, the error will not go away. Damage can be immense. Following this sort of event, it is necessary to stop the operation and resynchronize the controller and step motor.

**FIGURE 5-16** The servomotor is precise and robust with continuous feedback. It is used in many home automation projects. (*Photo courtesy of Mouser Electronics.*)

The servomotor incorporates a closed-loop control system. The step motor–controller combination is open loop. For the servomotor's closed-loop operation to work, a sensor is required. It should be located, if possible, at the load rather than at the motor in case the fault takes place between them. The servomotor control senses the position and speed of the motor or load. When an error is detected, the controller sends appropriate command data to the motor, and the system returns to normal.

Servomotors are not specific motor types, but they are characterized by the fact that there is a motor sensor-controller combination. The control is always closed loop. Most servomotors are capable of rotating 120 to 180 degrees. A continuous servomotor is able to turn 360 degrees in either direction. Any one of them is appropriate for robotic assembly-line production. Often a single robot will have several motors. There are numerous other applications, such as solar array tracking and the clock drive for an astronomical telescope, allowing it to remain pointed at a celestial object for long hours needed for time-exposure photographs. With a USB connection to a computer, these instruments can instantly find numerous deep-sky objects.

When we talk about ac motors, it is generally understood that we are referring to synchronous or asynchronous (induction) motors, both of which require an external ac power source. Thomas Edison envisioned a strictly dc electrical generating and distribution system. Beginning in lower Manhattan but quickly spreading far and wide, his system was an outstanding achievement and a tribute to his meticulous attention to detail and ability to think big. But George Westinghouse and Nikola Tesla saw the possibilities for a radically different electrical system that within a few years brought a new kind of electrical power into factories, businesses, and homes worldwide.

Dc power flows from the source to the load when a steady voltage causes current to flow through the circuit. At the source, there are two terminals, one positive and the other negative. If conductors are connected to these terminals and to a load, terminals at the load also will be positive and negative.

Electrons are negative and flow from the negative to the positive pole of a source such as a battery or dc generator. This is strictly a matter of semantics, however, and derives from the fact that originally it was thought that current flowed from positive to negative. Protons and neutrons attract one another because they have opposite charges, but there is nothing intrinsically positive about a proton and negative about an electron. It is just the way we designate them for historical reasons.

Ac also involves the motion of electrons. The difference is that the two poles reverse polarity either 100 or 120 times per second depending on the country. Two reversals constitute a *cycle*, so the most common frequencies are 50 and 60 Hz.

Ac does not switch abruptly. Square-wave power of that sort would have several disadvantages. The fast rise and fall times would make for powerful and harmful harmonics. Ac as supplied by utilities throughout the world consists of a pure sine wave, which is a consequence of the rotary nature of the generator. Figures 5-17 through 5-19 show sine-wave, square-wave, and triangular-wave graphic representations.

The voltage produced by a rotary generator, if plotted on a graph where the vertical axis represents electromotive force in volts and the horizontal axis represents time in seconds, conforms to a sine wave. There are several points worth mentioning: the peak-to-peak voltage is higher than the nominal line voltage, as

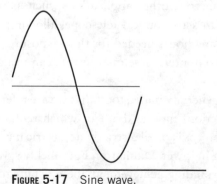

**Figure 5-17**  Sine wave.

**Figure 5-18**  Square wave.

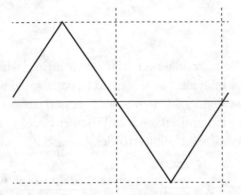

**Figure 5-19**    Triangular wave.

specified by the utility and required to run electrical equipment. What we are concerned with is the root-mean-square (RMS) value.

Looking at the sine waveform, notice that the rate of change is not the same at all times, as it would be in a triangular waveform. In a sine wave, the rate of change is greatest when the voltage is least, and the rate of change is least when the voltage is greatest. This fact is of great importance to electronics technicians, for whom capacitive and inductive loads are objects of intense interest. This is so because the impedance of these loads depends on the rate of change, not the actual value, of the applied voltage. In a capacitor, the greatest amount of current flows when the voltage is least, crossing the zero line.

For a purely capacitive or purely inductive load, the applied voltage and measured current are 90 degrees out of phase. In an inductive circuit, the current lags behind the voltage, whereas in a capacitive circuit, the current leads the voltage. In a resistive circuit, the waveforms of the applied voltage and measured current coincide and are said to be *in phase*.

## What Is Root Mean Square?

In the language of mathematics, the *root mean square* of a quantity, alternately called the *quadratic mean*, Is the square root of the arithmetic mean of the squares of all values, and it is used as a true measure of those numbers without regard to their sign. In electronics, the *RMS voltage* is the effective voltage as opposed to the peak-to-peak value, as seen in the waveform. In volts mode, your multimeter automatically displays the RMS value. This is the quantity that is meant at all times, unless you are considering the waveform as depicted on the screen of an oscilloscope.

## Why It Is Important

All of this is of limited interest to the home crafter-electrician when wiring switches and receptacles in an addition or new building, but it is good to understand what is going on for the perspective that it provides. Moreover, as discussed later in this book, when it comes to data networking projects such as Ethernet circuits, a bit of background knowledge will enhance your ability to design, install, troubleshoot, and repair such systems.

How is this relevant to motors? The introduction of ac made possible some new types of motors—synchronous and asynchronous (induction) and both of these in single- and three-phase versions. Ac motors had some advantages, but for many decades, the dc motor could not be matched when it came to easy reversibility and smooth speed control. Since the mid-twentieth century, with introduction of the variable-frequency drive (VFD), ac motors have become competitive, if not altogether dominant, in these areas as well.

One of the advantages of ac is that the utility-supplied power can provide external commutation so that switching is obtained without effort. In effect, the shaft of the utility's generator becomes the commutator for the motor.

The speed of a synchronous motor is exactly in step with an integral multiple of the utility line frequency. The multiplier is equal to the number of poles in the rotor. There are several configurations. Expect to see electromagnetic windings built into the stator. The line current is fed to these windings, and the magnetic field of the stator rotates in accord with the line-current pulses, and the rotor turns at that speed.

Synchronous motors come in many sizes, conforming to the intended use. Small synchronous motors are used as timers and clocks. The speed depends exactly on the line frequency, so they are very accurate. Utilities watch their frequency carefully, and if they detect an error, they temporarily increase or decrease the speed of the generator so as to get back on track.

Small synchronous motors will start on their own, so they are said to be *self-exciting*. Large synchronous motors have too much inertia at rest to be self-starting. An auxiliary induction motor (*pony motor*) is needed to get them going, or induction windings may be built right into the motor.

A large synchronous motor is very expensive to manufacture, but it is used in special applications where accurate synchronization to the line frequency is desired. Additionally, large synchronous motors are highly efficient, cheaper to run, and provide power correction in facilities where there is heavy inductive loading. Troubleshooting a synchronous motor is similar to troubleshooting a dc motor. After verifying that the supply voltage is present at the motor terminals, ring it out with an ohmmeter. Resistance between any (and all) windings and ground should be in the high-megaohm range, and resistance of each winding should be low but not zero.

## A Great Invention

Most ac motors in operation throughout the world today are single- and three-phase induction motors. They are relatively inexpensive to manufacture and, without brushes, for the most part maintenance-free. The small single-phase induction motor, as found in many stationary home tools and appliances, has sealed bearings. Because lubrication and brush maintenance are not required, the motor generally runs the life of the appliance with no maintenance or attention of any kind.

The induction motor was invented independently by Galileo Ferraris and Nikola Tesla in the 1880s, and working models were marketed first by Westinghouse and then by General Electric, eventually populating factory, farm, and home wherever electrical distribution lines permitted. As in synchronous motors, utility-supplied ac is connected to the induction motor's stator, producing a rotating magnetic field. What is radically different is the way in which the energy gets into the spinning rotor. Stator and rotor are, in effect, the primary and secondary of a transformer. The rotor is energized by magnetic induction, with no direct wiring. Then, as in other electric motors, the rotor's magnetic field chases the stator's magnetic field, and this is what makes the shaft turn.

Nevertheless, an induction motor is not synchronous. The rotor's spin depends on the ac line frequency, but spin and line frequency are not locked in synch. The rotor's spin is slower than the rotation of the stator's magnetic field by a certain amount, typically 10 percent. It is not merely out of phase like voltage and current waveforms when there is a reactive load. The rotor turns more slowly than the stator's magnetic field by an amount determined by the load and the torque of the motor. This amount is called *slip*, and it is always relevant to an induction motor. If the rotor's and stator's magnetic fields were, hypothetically, to turn at the same speed, there would be no inductive coupling and no reason for the rotor to rotate. Slip therefore is a necessary part of the induction-motor picture. It should not be regarded as some sort of wasted energy. Tesla, and later Charles Proteus Steinmetz at General Electric figured all this out and quantified it.

Tesla also conceived of three-phase power. It is more efficient to generate, distribute, and use than single-phase power once the infrastructure is in place. The

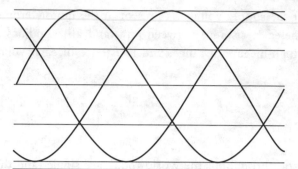

**FIGURE 5-20**    Three-phase waveform.

waveform is shown in Figure 5-20. (There is also a two-phase system, but it is rarely used.)

Most single-family dwellings have single-phase services, but if the size of the load requires a 400-amp or greater service, it probably will be three phase. An ordinary Delco-Remy automotive alternator generates three-phase power, which is converted to dc in a diode network that is inside the alternator.

The home crafter-electrician may never have occasion to wire a three-phase load, but you should know what is involved. Many individuals have acquired perfectly viable high-quality woodworking or other equipment for the home or shop, only to find that the item cannot be powered by the single-phase system without resorting to conversion strategies.

Actually, once you know the basics, three-phase wiring is easier than single-phase wiring. This is so because for a given size load, the conductors are smaller. In large-horsepower applications, this is critical.

A good way to get a feel for three-phase circuits is to travel the roads, looking at power lines, transformer arrays, and services. Have someone else drive, lest you become so intent on the circuitry that you drive into a power pole, knocking out the very system that you intended only to observe.

Large high-voltage transmission lines, unless they are dc, are always three phase. The three-phase conductors are invariably at the top. They may consist of three pairs. The two conductors that make up a pair can be close together. If by mishap they contact each other, it is not catastrophic because they are at the same potential. They are connected in parallel so as to permit the use of smaller and more manageable wires.

Experienced utility workers can tell at a glance the rated voltage of a line, judging by the height of the tower and construction of the insulators. If you follow the high-voltage power line, you will see that it terminates at a substation.

Large restaurants, hotels, and grocery stores with lots of refrigeration, as well as good-sized office buildings and most factories, have three-phase services. The ones that are aerial are easy to distinguish from single-phase services because they have three conductors (not counting the grounded neutral) or six if they have been paralleled.

A three-phase meter socket has three input and three output lugs, possibly twin lugs to accommodate paralleled conductors. The service-entrance conductors usually disappear into the building. Inside, they go into a three-phase entrance panel unless there is a main disconnect in a separate enclosure.

The three-phase entrance panel superficially resembles a single-phase entrance panel, but they are not the same. A three-phase entrance panel has a three-phase main breaker, shown in Figure 5-21 (or three separate cartridge fuses). The main breaker has three input lugs, and it energizes three vertical bus bars that run down almost to the bottom of the box, with a space below them so that wiring can cross over.

The main breaker has a single handle, permitting the three bus bars and all loads to be powered down simultaneously or, in the unlikely event that the main

**FIGURE 5-21**   A three-phase circuit breaker, suitable for supplying power to a small three-phase induction motor or similar load.

trips out, ensuring that all three open together. This type of panel will accept single-, double-, or triple-pole breakers for branch circuits and feeders. Thus three-phase loads or single-phase pole-to-pole or pole-to-ground loads may be connected.

A three-phase motor is powered through a three-pole breaker of appropriate ampere rating. Unless there is auxiliary equipment, the neutral is not required. Run the three hot legs and, of course, the equipment-grounding conductor in the same conduit or cable first to the motor controller and from there to the motor.

One of the interesting things about a three-phase motor is that the direction of rotation can be reversed simply by reversing the connections of any two of the three wires. This may be done at the output of the three-pole branch-circuit breaker in the entrance panel or load center, at the input or output at the motor controller, at the motor terminals, or at any junction box as long as other loads will not be adversely affected. Choose the location that is most accessible. To facilitate this operation, sufficient wire slack should be left in the initial installation.

The wiring for correct rotation may be determined by trial and error for some types of equipment, but care must be taken. Some types of pumps, if rotated in the wrong direction, are damaged to the extent that the seals must be replaced, involving a teardown. Also, a hazardous situation could be created. If there is a reversing switch, it will not matter how the rotation is established at the outset, provided that the switch is not labeled. A circular saw or a fan can be wired by trial and error, visually observing the rotation. A three-phase motor rotation meter, shown in Figure 5-22, is helpful in getting the rotation right.

Some pumps and blowers will push the liquid or gas in the correct direction regardless of direction of rotation, but the output will be greater one way than the other. This is so because the impeller blades are cupped, and they push more volume when the concave surface is leading. This is noticeably true for a three-phase submersible deep-well water pump. You can time how long it takes to fill a 5-gallon pail from a faucet, reversing connections at the control box or wellhead to find the best output.

Another consideration arises when wiring a three-phase motor. For maximum motor life and economy of operation, the three phases should be balanced. It is possible, especially in an older motor, that the windings have not retained their original impedances. Similarly, the line voltages for the three phases may differ by a slight amount. The idea is to match loads to lines in such a way that the errors cancel out rather than reinforce one another. The reconfiguration must be done so as to preserve correct motor rotation.

First, establish the desired motor rotation. Then, at a convenient splice location, label both line conductors and terminals A, B, and C. With the motor driving its load, use a clamp-on ammeter to measure the current in each leg. Write down

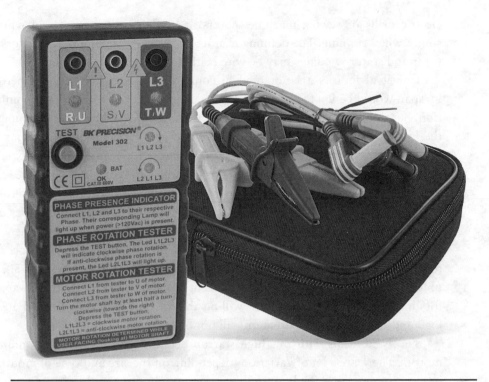

**Figure 5-22** A three-phase motor rotation meter. (*Photo courtesy of Mouser Electronics.*)

the readings. Then roll the connections, joining *A* to *B*, *B* to *C*, and *C* to *A*. This operation will not change the direction of motor rotation. Again, write down the three readings. Finally, roll the connections one more time, joining *A* to *C*, *B* to *A*, and *C* to *B*. Again record the readings. Choose the set of connections that gives the most uniform readings. We have to realize that the utility line voltage may vary depending on loading. The utility makes every effort to balance the load, but no one can foresee the use that customers will make of their accounts at all times. You may have to revisit the three-phase current readings at different times of day and night and on different days of the week to get the total picture.

## The Universal Motor: Small, Inexpensive, Ubiquitous

Another type of motor that is very common in the home and shop is the *universal motor*, so called because it can run on dc or ac or at other frequencies within limits. The universal motor is small, lightweight, and its speed and direction of rotation may be changed, making it ideal for many applications, such as portable electric drills or sewing machine motors. It makes a distinctive loud whirring sound when running. The defining characteristic is that unlike other motors, the rotor and stator windings may be wired either in series or in parallel.

Universal motors are not used in continuous-duty applications, where the lifespan would be limited because of the high-speed commutator-brush combination. But the ease of control by means of internal speed and reverse switching and the high speed and torque relative to size and weight make them ideal for hair dryers, electric drills, food mixers, and many other applications in the home and shop. Brushes are easy to change, and this may have to be done two or three times in the life of the motor. Because many of these tools and portable appliances are moved around a lot while in use, a frequent cause of failure is the power cord, which can be replaced if you can save the old strain relief or fabricate a new one.

Following this brief survey of the types of motors that you may find occasion to rewire or repair in the home or shop, we'll return to *NEC* Article 430. The object is to learn how to size the overcurrent device(s) and branch-circuit conductors in such a way that the motor will be able to start and run while driving the load and yet a hazardous situation will not be created.

The *NEC* deals with this high startup demand in a way that is unique to motors. There are a lot of variations depending on the size and type of motor, but in most applications, the bottom line is that you do *not* size the circuit according to the full-load current on the nameplate. Instead, you take the horsepower off the nameplate and refer to the appropriate table at the end of Article 430. This table gives a different full-load current. The circuit is sized accordingly. Other tables in Article 430 permit choosing a much larger branch-circuit overcurrent device than would normally correspond to the circuit size. The key concept, and it may seem counterintuitive, is that this overcurrent device is intended to protect the branch-circuit conductors only from short-circuit and ground-fault events but not from overloads.

If the motor draws excessive current because of internal winding insulation breakdown, rotor lockup, or stalling due to the driven load, the circuit will be interrupted by a separate overload device that is in a motor controller adjacent to the motor or in the motor. This slow-acting overcurrent device will protect the motor and supply circuit in the event of overload but will not prevent the motor from starting up and attaining rated speed, whereupon the counter-electromotive force will limit the current.

Apprentice electricians for many years have struggled with this arrangement, fearing that the branch-circuit conductors will overheat and incinerate the building. But it is a fine-tuned, carefully engineered design that has proven safe and reliable when constructed according to the *NEC* mandates contained in Article

430. Any time you have occasion to wire a motor that is larger than a cord-and-plug–connected fractional-horsepower unit, it is time to go through Article 430 carefully and follow the requirements that pertain to your installation.

## *NEC* Chapter 5: "Special Occupancies"

Chapter 5 opens with extensive coverage of hazardous conditions, most of which is not relevant to the home crafter-electrician, with some important exceptions that we shall note. Special designs and installation techniques are necessary in hazardous locations, places where there are or may at times be accumulations of explosive or flammable gases, volatile liquids, dusts, or larger fibers that can become airborne. These locations are identified and delineated as three separate classes depending on the flammable material. Each class is divided into two divisions corresponding to the immediacy of the hazard. So there are six types of locations, Class I, Division 1 being the most hazardous and Class III, Division 2 being the least hazardous. Areas that are not hazardous, such as most residential locations, are termed *unclassified*. The six classified locations require different types of wiring and electrical equipment that are approved for the location, such as that shown in Figure 5-23.

**Figure 5-23**  Classified locations require specialized materials and wiring methods.

NEC Chapter 5 also covers other special locations that, while not hazardous in the sense of the classified locations mentioned earlier, are sensitive in varying degrees and require wiring methods and materials that go beyond the ordinary Romex installation that is seen in a low-rise single-family home. The home crafter-electrician must be sophisticated and vigilant regarding any of these issues that may arise. Here are some of the articles that cover special locations that may be relevant to a home wiring project:

- Article 511, "Commercial Garages, Repair and Storage"
- Article 514, "Motor Fuel Dispensing Facilities"
- Article 516, "Spray Application, Dipping, Coating, and Printing Processes Using Flammable Combustible Materials"
- Article 550, "Mobile Homes, Manufactured Homes, and Mobile Home Parks"
- Article 590, "Temporary Installations"

You may wonder what Article 511, "Commercial Garages," has to do with a home wiring project. A good proportion of residential properties include a garage that is attached to or detached from the primary house. If the garage is attached, it must be wired and must conform to the NEC. If it is detached, power is optional, but if it is wired, it must comply. The garage can have its own service, or it can be powered via underground or aerial feeder from the house. How it is wired depends on whether the garage is deemed to be residential or commercial. A residential garage can be wired in Romex, and the wiring resembles ordinary house wiring with GFCI receptacles. If the garage is commercial, you're looking at metal raceways.

A great many commercial garages are owner-built, and it is a common Code violation to see them wired in Romex with no thought of the hazards. There are gray areas. Often the garage is used by a backyard mechanic who works on automotive vehicles on a part-time basis. The garage may begin as a strictly residential parking garage, but as the teenager in the house grows older, it may become the site of extensive automotive repair and fabrication. The combination of heat systems, wet floors, welding equipment, grinders, battery chargers, and disassembled fuel systems opens the possibility for shock and fire hazards.

Many of these backyard facilities are as much or more prone to disaster than a full-scale commercial garage with a large payroll. Before designing and installing a wiring system for one of these buildings, a decision should be made as to whether it is to be considered residential or commercial. Not only the immediate usage but future activities need to be considered. The key concept is not at all whether the work is to be done for pay. It is based on the presence of vehicles or equipment that use flammable fuels (including propane or natural gas) and whether the fuel systems will be opened, allowing for fuel vapors to become a factor.

Do you have a gas pump on your property? Gasoline is a highly volatile liquid that under certain conditions releases explosive vapors. You don't want to just throw together an electrically operated gas pump, as seen on many farm properties. Consult *NEC* Article 514 and observe all setbacks and wiring standards.

If you have a woodworking shop, parts of it may be Class II, Division 1 or 2. This depends on the volume of sawdust, how fine it is, whether a collection system is in place, how large and well ventilated the area is, and other factors. Finely divided sawdust can remain suspended in the air, and at a certain concentrations, it becomes a potential explosion or flash-fire hazard, awaiting an electrical arc from a switch, motor-controller relay, or brushes within a motor. Over a period of time, standard wall boxes can become packed with sawdust, preventing normal heat dissipation. Sawdust can settle on a motor enclosure, especially if there is an oily film. This will cause the motor to overheat, setting the stage for a smoldering fire that can take off after the shop is closed for the night.

Associated with a woodworking shop is sometimes spray-painting activity. This can involve a full-scale spray booth or merely an area where occasional airborne paint or solvent is present. Refer to *NEC* Article 516.

If you have a full-scale spray booth, an exhaust fan, lighting with a switch, and receptacles may be needed. Because airborne solvents and paints will be present at times, the area is classified. The safest and least-expensive way to deal with this is to keep all electrical wiring and equipment out of the area. The exhaust fan can be located under a shroud outside, high above grade. Lighting can be outside the spray booth, separated by glass.

If the building is a mobile home, it will be prewired during manufacture. Additions and alterations to mobile home wiring are notoriously difficult because the breaker box is recessed inside a finish wall, usually in a tight closet-like space that serves as a utility room. Wall finish is fastened to undersized studs by means of adhesive. The box will be small, and often there is no space for an additional circuit. Because everything inside is finished, it may be necessary to resort to surgical techniques. Also, because the raceway that brings the feeder into the box from under the mobile home is technically not part of the service, it is permissible to pull in an extra branch circuit, provided that the conduit fill limit is observed.

The home crafter-electrician or a local professional will have to build the service. Mobile homes are different from other buildings in this respect. An aerial service with meter is never attached to a mobile home. Instead, the service drop or lateral is brought to a separate pedestal at a distance from the mobile home that is specified by the utility. This pedestal contains the meter and the main disconnect. The conductors between the main disconnect and the mobile home constitute a feeder, with overcurrent protection at both ends. It is a two-leg 240-volt circuit with a neutral and equipment-grounding conductor, four wires in all. It is brought underground, usually in PVC conduit, if possible, to a location directly under the breaker box. The best procedure is to have this feeder in place before the mobile home is delivered to the site.

## Temporary Installations

Most large electrical projects require some temporary wiring, just to provide power so that the main project can be done. Article 590 contains requirements for temporary installations. Some of these requirements are less stringent than for permanent work, but this is no license for a chaotic, low-quality product. Some workers maintain a very relaxed stance toward work-site wiring, a "get 'er done" attitude predominating. Actually, great care should be taken to ensure that proper grounding, protection of conductors, proactive elimination of potential fire hazards, and other good procedures are observed. A very dangerous practice is sawing off extension-cord ground plugs because continuity of the equipment ground is a definite safety issue for prevention of hazardous voltage on normally non-current-carrying conductive materials. No one wants to carry the memory of a job-site fatality. To prevent this, compliant temporary wiring is the place to start.

Temporary wiring is permitted during a period of construction, remodeling, maintenance, repair, or demolition of buildings, structures, equipment, or similar activities. It is to be removed immediately on completion. It is permitted for up to 90 days for holiday decorative lighting. Except as specifically modified in Article 590, all other *NEC* requirements for permanent installations apply to temporary wiring installations.

Romex is permitted as temporary wiring in all locations. On construction sites, unlike in permanent wiring, enclosures are not required for wire-nut splices. Ground continuity of the equipment-grounding conductor must be maintained. GFCI protection is required for all temporary receptacles. Cord sets for portable use are permitted to meet this requirement.

## Special Equipment

*NEC* Chapter 6, "Special Equipment," contains a series of articles each of which focuses on a type of equipment that requires specialized wiring designs and installation techniques. Most of this equipment falls outside of the home crafter-electrician's primary field of interest, but some of it may find its way into a residence, and we shall take note of it.

Article 620, "Elevators, Dumbwaiters, Escalators, Moving Walks, Platform Lifts, and Stairway Lifts," is a highly technical article that covers this type of work. It is unusual for the home crafter-electrician to wire an elevator, much less a bank of elevators, in the ordinary low-rise residential occupancy. But it is a distinct possibility that a stairway chairlift may be installed in the home for the benefit of a disabled person when the bedroom is not on the first floor.

A stairway lift is purchased as a complete assembly with a rail that is sized to fit the existing stairway in the home, straight or curved. Along with the purchase, professional installation is available. A skilled homeowner certainly can do this work, including the electrical component. Full documentation that comes with the model should answer any question that arises.

## *NEC* Article 625, "Electric Vehicle Charging Systems and Welders"

As hybrid and fully electric vehicles become more common, homes are being equipped with vehicle charging systems. Such a system can be connected to the utility service, a dedicated generator, fuel cells, or wind or solar power. These sources can be integrated with one another in various permutations for increased reliability and/or economy. The *NEC* devotes an article to equipment requirements, and as always, the focus is safety. To this end, there are mandates that pertain to the electric vehicle coupler, grounding, GFCI placement, ventilation, disconnecting means, and interactive systems. If you are setting up an electric vehicle charging system, consult Article 625 to make sure that you get all the details right.

Article 630, "Electric Welders," provides guidance on wiring provisions for this type of equipment. It is increasingly finding its way into home workshops. If run through the home electrical service, it may be necessary to upgrade to 200 amperes. A common type of machine is the 225-ampere stick welder. The ampere figure refers to the low-voltage output, not the power supply required. Larger dc welders give superb performance and work well with many difficult metals. And then there are the smaller 120-volt shielded-arc wire-feed models, excellent for light to medium metal fabrication and repair.

If there is a unit on-off switch on the welder, and if it is cord-and-plug connected and within sight of the branch-circuit breaker, there is no need for a separate disconnect. The only task is determining the correct overcurrent device and branch-circuit conductor size.

Welders have a very forgiving duty cycle. Welding is typically a stop-and-go operation, with frequent breaks for chipping, grinding, and repositioning the work. Additionally, the waveform of the welder output current has an inherent duty cycle, which is marked on the welder. It differs for the various heats. (One of the heats, typically 75 amperes, indicated by a circle around the number, has a heavier winding and is used for thawing frozen underground metal pipes, which may require hours of continuous operation.)

Assuming that you are not doing a group installation, the ampacity of the supply conductors (not to be confused with the welding cable) is not to be less than the $I1_{eff}$ value on the rating plate attached to the welder. This includes the duty-cycle allowance. Alternatively, it is permissible to multiply the rated primary current by the multiplier given in *NEC* Table 630.11(A). Notice that these duty cycles run quite low, permitting small supply conductors.

Several 240-volt receptacles of the proper ampere rating can be daisy-chained around the garage or shop, permitting the welder to be operated at different locations. A receptacle should be placed near the door so that the welder can be positioned outdoors if needed. An outdoor receptacle on the exterior wall of the building is also a possibility.

## A Big Wiring Project

*NEC* Article 680, "Swimming Pools, Fountains, and Similar Installations," is not too many pages long, but it is densely written and contains some highly technical information. Around bodies of water, there is a high potential for shock. But the provisions in this article, if followed faithfully, will mitigate that hazard. What is needed is great attention to detail and careful workmanship.

The provisions of Article 680 apply to construction and installation of electrical wiring for and equipment in or adjacent to all swimming, wading, therapeutic, and decorative pools, fountains, hot tubs, spas, and hydromassage bathtubs, whether permanently installed or storable, and for metallic auxiliary equipment, such as pumps, filters, and similar equipment. The statement of scope is quite broad. Where there is electricity and water, whether human-made or natural, you've got some very technical and exacting work to do. Fortunately, it is all spelled out in Article 680, so it is just a question of going through this article until all the details of the installation are grasped.

It is beyond the scope of this book to cover all the details of the electrical part of a swimming pool installation. For now, we'll indicate some of the main areas of concern. The good news for the home crafter-electrician is that you don't have to become an expert in this broad subject until such time as you are called on to do such an installation. If this does happen, a wise choice would be to call in a professional electrician on a consultative basis. In the meantime, it is good to delve into the subject just to find out what is involved.

Naturally, there are all sorts of clearances. Service-drop conductors, communications, radio and television coaxial cables, network-powered broadband communications systems, and other overhead conductors that pass near the pool are required to maintain horizontal distances from the inside wall of the pool and water edge. This is in addition to ground clearances that are in effect even if there were no pool.

Underground wiring, depending on raceway type, must meet prescribed minimum cover requirements. Receptacles within 20 feet of the inside wall of a pool must be GFCI protected. Luminaires have required vertical clearances from the pool depending on whether they are indoors or outside. Switches are to be at least 5 feet horizontally from the inside walls of a pool unless separated by a solid fence, wall, or other permanent barrier.

All of this is straightforward. Where swimming pool wiring differs from other types of electrical wiring in the home is the whole subject of bonding, and this becomes complex and exacting. To comply with *NEC* requirements for swimming pool wiring, it is necessary to develop a detailed plan in advance because of the need for equipotential bonding. If the swimming pool consists in part of below-grade concrete, it will have to incorporate a rebar grid that is bonded to the electrical grounding system. This is also true of slablike construction that surrounds the perimeter of the pool.

If there is an electrical fault, steep voltage gradients can become established within the pool and in the area surrounding it. If a person, especially with bare feet, walks on the ground or occupies the pool and is positioned perpendicular to the voltage gradient, severe shock and electrocution are possibilities. To eliminate this voltage gradient, the *NEC* requires equipotential bonding. This has to be set up as part of the formwork before the concrete is poured.

The bonding is accomplished by using insulated or bare solid (not stranded) 8 AWG or larger conductors connected to all the following:

- Conductive pool shells
- Perimeter surfaces
- Metallic components
- Underwater lighting

- Metal fittings
- Electrical equipment
- Pool water

This 8 AWG bonding conductor need not be run all the way back to the service-entrance panel. It is only necessary for it to connect to the electrical grounding system. Under no circumstances should a separate floating-ground electrode be thought to suffice because this would be totally ineffective. Article 680 goes on to cover, as separate topics, spas and hot tubs and fountains. Here are important grounding, bonding, GFCI, and other requirements that must be met, so careful review of the article is essential before beginning the installation.

Article 682, "Natural and Artificially Made Bodies of Water," deals with some of these same issues, such as the need for an equipotential plane and bonding.

Article 690, "Solar Photovoltaic Systems," Article 692, "Fuel Cell Systems," and Article 694, "Wind Electric Systems" include *NEC* requirements for these interesting and increasingly timely technologies. We'll discuss details in Chapter 14.

These are the 2014 *NEC* articles that are most relevant to residential electrical installations. A home crafter-electrician who is serious about doing quality work that will be safe for family members, visitors, and future owners, will obtain this volume and adhere to its mandates.

## Some Special Wiring Methods

It's not all Romex! Many cable and raceway materials are suitable for various applications. Here's a brief summary.

### Rigid Metal Conduit

Rigid metal conduit (Type RMC) is the heaviest of the metallic raceways. It is similar to galvanized water pipe in that the threads are compatible. But water pipe must never be used in electrical work. The inside diameter is less and the inside surface is rougher, so conductors pull harder and may be damaged. RMC is suitable for the most demanding applications, and it may be used anywhere that raceway is permitted, including all home projects. For underground lines, it is useful where it is impossible to go deep due to bedrock. RMC is expensive and difficult to thread and bend, so it is rarely used where other raceways are permitted. RMC is to be secured within 3 feet of each outlet box, junction box, device box, cabinet, conduit body, or other conduit termination. RMC is to be supported at intervals not exceeding 10 feet.

## Liquidtight Flexible Metal Conduit

Liquidtight flexible metal conduit (Type LFMC) is similar to FMC, but it has a liquidtight outer plastic covering. Securing and supporting requirements are the same as for FMC. LFMC is permitted as an equipment-grounding conductor. Its use is permitted outdoors, for direct burial, and embedded in concrete.

## Liquidtight Flexible Nonmetallic Conduit

Liquidtight flexible nonmetallic conduit (Type LFNC) is similar in appearance and use to Type LFMC, but it does not have the metal sheath inside the outer plastic covering. It is less expensive and may be more prone to physical damage, but otherwise it is suitable for many applications.

## Rigid Polyvinyl Chloride Conduit

Rigid polyvinyl chloride conduit (Type PVC) is permitted concealed and exposed in indoor and outdoor locations. For indoor raceways, Type EMT is preferable because it has a more finished appearance. Long horizontal runs of Type PVC are to be avoided because it is prone to sagging and buckling when exposed to changes in temperature. Type PVC is relatively inexpensive as raceways go, and it may be used to good effect for underground electrical lines and services and embedded in concrete.

## Electrical Metallic Tubing

When it comes to raceways, electrical metallic tubing (Type EMT), shown in Figure 5-24, is the workhorse in many electrical installations. Technically not a conduit but rather a tubing, this metal raceway is suitable for all but the harshest environments. It is easier to cut and bend than Type RMC, and it is less expensive, so it is quite user friendly. Type EMT may be deployed exposed or concealed, indoors or outside. Where used in wet areas, compression (rather than set screw) fittings are needed. All cut ends of Type EMT are to be reamed. It is not threaded on the work site. Type EMT is to be supported at intervals not exceeding 10 feet, and it is to be securely supported within 3 feet of each outlet box, junction box, device box, cabinet body, or other tubing termination. Type EMT is permitted as an equipment-grounding conductor, but better electricians pull an additional green wire. Type EMT is not used much in underground installations or embedded in concrete because the less expensive PVC is preferred in those applications.

**Figure 5-24**  Type EMT is an excellent general-purpose metal raceway for indoor or outdoor use.

## Flexible Metal Conduit

Flexible metal conduit (Type FMC), shown in Figure 5-25, is a convenient multi-purpose flexible conduit that is permitted in residential applications except wet locations, underground, or embedded in poured concrete. All cut ends must be trimmed or otherwise finished to remove rough edges, except where fittings that thread into the convolution are used. Type FMC is to be securely fastened in place within 12 inches of each box, cabinet, conduit body, or other termination, and it must be supported and secured at intervals not exceeding 4½ feet. Type FMC qualifies as an equipment-grounding conductor within limitations, but the best policy is to pull a separate green conductor for the purpose.

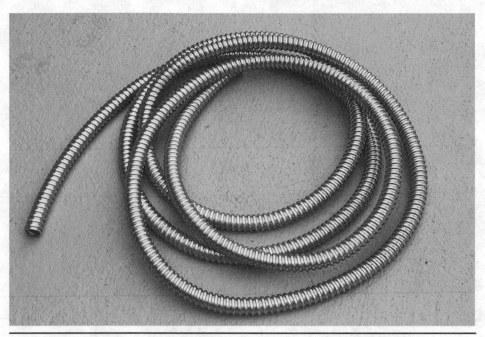

**FIGURE 5-25**   Type FMC provides excellent protection where flexibility is needed.

# CHAPTER 6

# Tools for Quality Electrical Work

The quality of your electrical installations will depend on your knowledge and expertise plus the degree of manual dexterity you have. Still another very significant part of the picture is the variety and quality of the tools that you bring to the job. To do successful electrical installations, the home crafter-electrician, no less than the full-time professional, must be well equipped. The tools that you will need fall into two broad categories—test equipment and installation tools.

## Test Equipment

We'll start with some inexpensive and modest items and progress toward the top of the line. The first and most basic need is to test for the presence or absence of voltage without regard to the exact level, where level is not an issue. You need an instrument that will do this to verify that wires and terminals are not live prior to working on them. The same tool is used when diagnosing and repairing equipment and circuits to determine whether they are energized. It is to be emphasized that the instrument should be tested immediately before each use to ensure that you are not deceived by a dead battery, open probes, or a burn-out indicator light or meter.

A very basic tool you can make at low cost is a plug-in lamp socket equipped with an appliance bulb. (These bulbs, because of the tight radius of the glass envelope, do not break easily, and the short filaments will withstand a lot of vibration

**Figure 6-1**    Plug-in lamp socket.

and rough handling.) This test light, shown in Figure 6-1, can be quickly plugged into any receptacle and left in place to indicate the status of the circuit while you work on it. For testing receptacles, it is superior to the best meter money can buy because with meter probes in a receptacle, you never know for sure when you are getting a good contact.

A variation on this type of indicator is an appliance bulb screwed into a lamp socket that is equipped with wire leads, as shown in Figure 6-2. Such a socket can be bought from your electrical supplier or saved from a light fixture that is being discarded. (It is good to have a dozen or so of these sockets with leads on hand because they are also used as temporary lighting after the branch circuits are roughed in but before the finish ceiling and wall materials have been installed.)

Moving up the cost scale, the next test instrument is the neon test light shown in Figure 6-3. It is quick and convenient for checking for presence or absence of voltage, and it rarely breaks or fails electrically. The bulb and probes are very

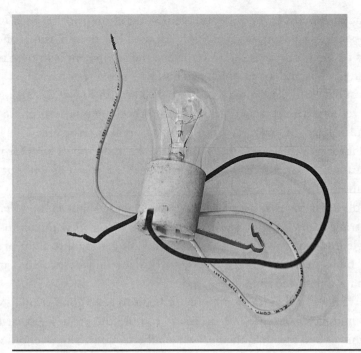

**Figure 6-2**   Lamp socket with leads.

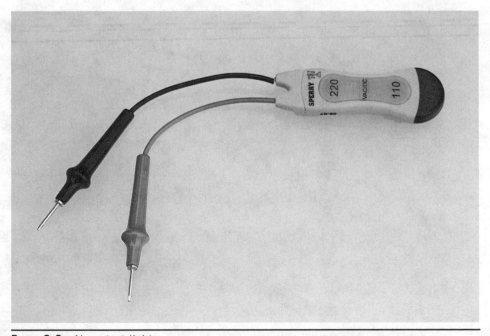

**Figure 6-3**   Neon test light.

durable, and the initial cost is modest. Still, verify operation prior to each use if there is any possibility that you will make contact with live terminals or conductors. The rule is to check first, and then, even when the equipment is known to be dead, use insulated tools and avoid touching conductors and terminals.

The neon test light will discriminate between 120 and 240 volts. You can rapidly move all over an entrance panel or load center, checking phase-to-phase and phase-to-ground voltages at main and branch-circuit breaker terminals. This little tester is excellent for working inside a tool or appliance and for checking for voltage at motor terminals. You also can slide a probe inside a wire nut without disrupting the connection.

Another tool that is favored by many experienced electricians is the solenoid voltage tester (trade name Wiggie), as shown in Figure 6-4. This compact instrument has flexible leads so that the probes are capable of getting into tight places. Multiple neon bulbs indicate voltage levels, so this instrument provides more detailed information than the simple neon light.

You may have noticed that it is sometimes difficult to get a good reading with a conventional mutimeter because it is necessary to hold the probes so that they contact the terminals and to simultaneously watch the readout. The solenoid voltage meter addresses this difficulty with great competence. At the top of the case are brackets that will hold one or both probes. With a probe mounted to the case,

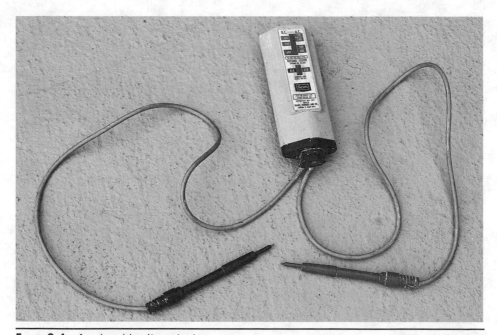

**Figure 6-4**  A solenoid voltage tester.

you can use one hand to position the other probe and, at the same time, hold the meter and view the readout. Moreover, when the probes contact metal surfaces that are at different voltage levels and when they break contact, the instrument emits a very loud click. As long as the probes are connected to an alternating-current (ac) circuit, there is a loud buzzing sound. This can be felt when you are holding the housing as a very distinct vibration, which is helpful in noisy surroundings.

The solenoid tester is line powered, so it does not depend on a battery, although some models require a battery for an auxiliary ohmmeter function. The instrument is a low-impedance voltmeter, so it should not be used for sensitive electronics work, where it would load the circuit. It is at its best for power and light circuits in ordinary house wiring. Because of

## High-Impedance versus Low-Impedance Voltmeter and Milliammeter

A multimeter in volts mode takes a valid reading when the probes are touched to the two electrical supply leads or across any of the loads. It is not necessary to cut open the circuit. If the voltage source has high impedance, such as in sensitive electronic equipment, a low-impedance meter such as a solenoid tester (Wiggy) should not be used because it will load the circuit and give an invalid reading. A high-impedance transistorized multimeter is better because it draws negligible current, does not load the circuit, and provides a valid reading. A clamp-on ammeter has no noticeable effect on the circuit. This is true of the in-line multimeter in milliameter mode. Because it is connected in series in the circuit to be tested, due to its very low impedance, it does not affect the circuit.

the low impedance, it should not be continuously connected to a live circuit for any length of time because it would overheat and burn out.

A side benefit is that the solenoid voltmeter can be used to test a ground-fault circuit interrupter (GFCI). Connect the probes to the hot line and equipment ground at the GFCI output or anywhere down the GFCI-protected line. The solenoid voltage tester draws enough current so that the GFCI will think that there is a ground fault and will trip out.

The electrician's next test instrument moving up in complexity and functionality is the multimeter, variously called a *volt-ohm milliammeter* (VOM) and a *digital volt milliammeter* (DVM). The multimeter is available in two versions, analog and digital. The old analog type is still manufactured and sold, and some old-time electricians prefer it. It has a needle that sweeps across a face on which are printed separate scales for different functions and ranges. A built-in reflective zone is help-

**Figure 6-5**    Analog multimeter.

ful in ensuring that the needle and scale are being viewed straight on. The analog meter, shown in Figure 6-5, is more reliable in very cold surroundings.

Today, most electricians favor the digital multimeter. An alphanumeric read-out displays the circuit parameter of interest. The multimeter is easy to read and interpret.

Digital multimeters are available in inexpensive versions for under $10, as shown in Figure 6-6, and this is a good way to get started. Better digital multimeters have more extensive ranges, are easier to read, and, with rugged rubberized cases, withstand dropping and rough handling. You'll pay over $100 for an excellent model with capacitor and diode check functions.

The voltage function is useful for checking for the presence or absence of voltage and will measure the exact level accurately and reliably. Although voltage measurements are made with the circuit powered up, the ohmmeter function is used only when the circuit is not live and any stored voltage (due to electrolytic

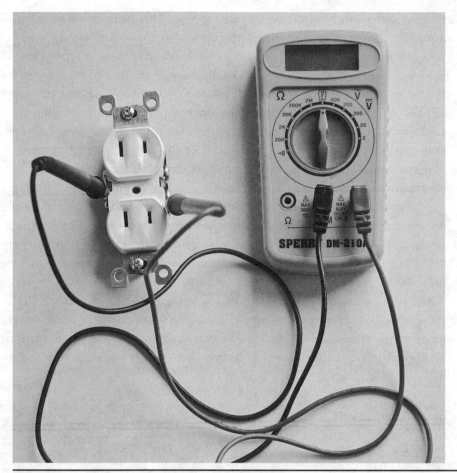

**FIGURE 6-6**    An inexpensive digital multimeter adequate for most electrical work.

capacitors or distributed capacitance) is bled out of the equipment. If there is a chance of a parallel current path, the circuit or device being tested has to be taken out of its larger electrical environment. To do this, it is necessary to open just one, not both, of the connections.

Most multimeters in the ohms function have a continuity indicator, which is an audio tone that sounds when the meter reads less than (typically) 30 ohms. It is very helpful for quickly testing circuits, equipment, and devices such as fuses and switches on a go, no-go basis.

Electric current in a circuit also can be measured, and this is done, of course, when the unit is powered up. There are two drawbacks to using a multimeter for this purpose. One is that to take an ampere measurement, it is necessary for the entire current to pass through the meter. Therefore, the circuit must be cut open so that the meter can be put in series with the load. The other disadvantage in using a multimeter for amperage measurements is that the meter is only able to deal with a low level of current. The amount of current drawn by a typical household load, even a low-wattage light bulb, would be outside the highest range of a multimeter in amperes mode and would instantly overheat and destroy the meter.

The multimeter is nevertheless the single most useful test instrument in an electrician's tool box, and the home crafter-electrician definitely must have one, even if it is an old yard sale analog meter or the low-end model purchased at a "big box" store. If you can justify a really good meter, that is the way to go. Doing electrical work, you'll use it at least once a day, and it will last many years.

Another piece of test equipment that professionals value highly is the clamp-on ammeter (trade name Amprobe), as shown in Figure 6-7. The home crafter-electrician will not need one of these instruments in the course of ordinary electrical installations in the home, but in a typical appliance repair shop, it may be the only way to get the information you need to proceed.

The clamp-on ammeter may be analog or digital. The old analog tester is line powered and has no battery. These vintage models work fine. The digital clamp-on ammeter, with an auxiliary ohms function, requires a battery and incorporates additional features such as a hold button that enables the readout to retain the highest reading.

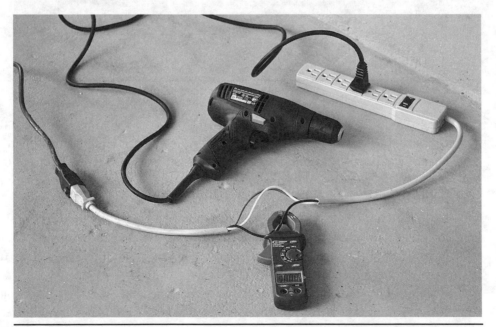

FIGURE 6-7  Clamp-on ammeter measuring current drawn by a portable power tool. Notice the homemade splitter, permitting reading a single conductor.

To measure current, you open the jaws and clamp them around a live (insulated!) conductor. The clamp-on ammeter is not placed in series with the load. Instead, it measures the magnetic field that surrounds any conductor through which current is passing. The instrument is capable of reading up to 200 amperes and has no measurable effect on the circuit being tested. It does not matter if the conductor is precisely centered in the jaws, nor does it matter if the conductor passes through at an angle.

The current drawn by a cord-and-plug–connected tool or appliance cannot be measured by clamping around the power cord. Because current in hot and neutral conductors is traveling in opposite directions, the induced magnetic fields cancel out, and a zero reading results. To measure current in a power cord, go inside the equipment, junction box, or entrance panel, where, with no simultaneous load online, it is possible to clamp around a single conductor, hot or neutral, to obtain a reading.

The circuit analyzer is an inexpensive test instrument that every electrician, professional or otherwise, should have. Plug it into a receptacle, and it will indicate any reversed or missing circuit elements. It is a quick and easy test for the presence or absence of voltage.

These are the basic electrical test instruments. For the home crafter-electrician, the entire array is not necessary. The instruments can be obtained on an as-needed basis. A viable plan would be to start with a neon test light, low-end multimeter, and circuit analyzer and take it from there.

## Electrical Installation Tools

As for installation tools, you will need the usual carpentry tools, such as a hammer, tape measure, plumb bob, level, hand saw, portable circular saw, electric drill with a good selection of bits, rubber mallet, and so on. You will also need the usual mechanic's tools, such as open-end and box wrenches; socket wrenches with $\frac{1}{4}$-, $\frac{3}{8}$-, and $\frac{1}{2}$-inch drives; adjustable wrenches; Vise-Grips; water-pump pliers (small, medium, and large); and slip-joint pliers.

There are many electrician's tools, some essential and others more expensive and in the optional category. Here's a rundown, beginning with some tools you can't be without.

You can make do with an automotive-type wire stripper, but the specialized electrician's wire stripper, shown in Figure 6-8, works much better. It is smaller, able to get into tight places, and incorporates a high-leverage, sharp wire cutter. If you value the way it snips easily through large copper conductors, you will refrain

**Figure 6-8** Professional electrician's wire stripper.

from using it on steel wire. The stripping cutters close down on the insulation without nicking the copper. There are different sizes marked for stripping solid versus stranded wire. The only disadvantage with this otherwise convenient little tool is that you have to look carefully to find the right cutter every time you want to strip a wire. The solution is to enamel small, different-color dots for stranded and solid 12 American Wire Gauge (AWG) wire. Then they, along with the neighboring 10 and 14 AWG strippers, will be easy to find.

The electrician's wire stripper also incorporates a gripper with knurled jaws that is useful for pulling wires out of tight places and myriad other tasks. The drilled holes are intended for determining conductor gauges, but they are useful also for forming wire ends so that they can be terminated.

Needle-nose pliers, shown in Figure 6-9, are absolutely essential for wiring switches and receptacles, as well as for many other tasks around the work site. You want a medium-sized pliers—not too large and not too small. The idea is that the diameter of one of the grippers should be equal to the diameter of the wire coil that will easily fit under a switch or receptacle screw terminal. If you form these wire ends properly, your terminations will be quick and easy with high-quality results. Grab the stripped end with your needle-nose pliers, and roll it in the proper direction (clockwise) so that when the terminal screw is tightened, the coil will wind tighter, not loosen. While coiling the wire end, form it at a slight angle so

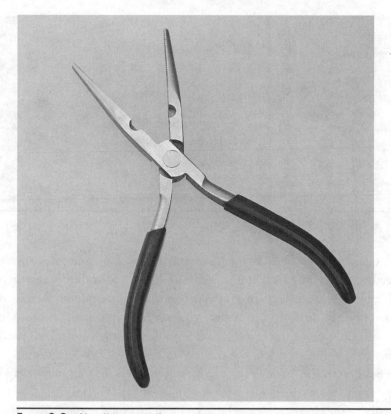

**FIGURE 6-9**    Needle-nose pliers.

that when you insert it under the screw, the leading end will lean in toward the device. This makes it easier to get the wire under the screw. The *NEC* states that such a coil is to form at least three-quarters of a circle around the screw. You can do better than that. Before tightening the screw, use your needle-nose pliers to squeeze the wire more tightly around the screw. Then tighten the screw very firmly, and you will have an excellent low-impedance connection that will last forever. Do this on each and every termination that you make.

As for screwdrivers, the very inexpensive six-way screwdriver with two Phillips sizes, two straight-blade sizes, and two nut drivers is very convenient for lots of work. The thick handle permits application of torque where needed. This tool will be adequate most of the time, but there are some specialized screwdrivers that are indispensible now and then.

The so-called electrician's screwdriver, shown in Figure 6-10, is a great help for certain difficult tasks. It has a very long, thin shank with a small, straight blade, and it is perfect for getting into tight places. With the thin shank, you can spin it rapidly between your fingers, making short work of a long haul.

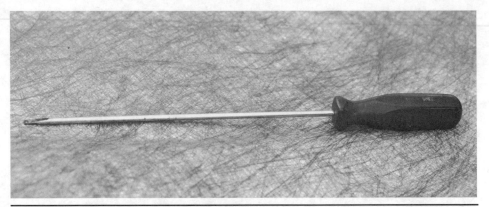

**FIGURE 6-10**    Electrician's screwdriver.

In close quarters, a standard screwdriver may be too long and possibly won't fit. You need to have on hand Phillips and straight-blade stubby screwdrivers, shown in Figure 6-11, in at least two different sizes. Sometimes a screw has to be started in a very tight, difficult place where you can't get your hand in position to start it and you can't maneuver it into position with a screwdriver with a magnetized bit. For this, you need a screw-starter type of screwdriver. You should have two or three different sizes. The best place to buy this type of tool may be a local auto parts store.

A necessity in the electrician's toolbox is a large diagonal cutter (*dike*), shown in Figure 6-12. It's perfect for cutting Romex, reaming electrical metallic tubing

**FIGURE 6-11**    Stubby screwdriver.

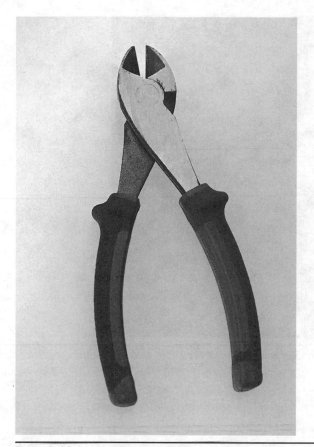

**FIGURE 6-12**    Large diagonal cutter.

(EMT), and many other tasks every day. Another good tool is a large pair of lineman's pliers. These are also used every day.

A cordless drill, shown in Figure 6-13, is a very basic tool that is used very frequently in electrical work. Besides drill bits, you will want Phillips bits for driving the ubiquitous drywall screws that are so much a part of our lives these days.

There are some more advanced electrician's tools that are good to have, but they are not essential for residential work. Many of them have to do with metal conduit installation. This is permitted in dwellings but is rarely seen because Romex is much less expensive and easier to install. In dwellings, raceways may be used in the service and for some electrical equipment. Polyvinyl chloride (PVC) works for most of these applications, so elaborate tooling is not needed.

Conduit benders, knockout punches, hole saws, and similar items can be purchased as needed, so it is not necessary to invest heavily in these types of specialized tools.

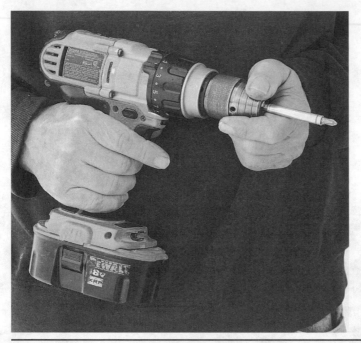

**FIGURE 6-13** It is hard to imagine doing much electrical work without a cordless drill.

# CHAPTER 7

# Techniques for Quality Electrical Work

A long with a good selection of quality tools, a good selection of quality techniques is essential for successful electrical projects. Some of this is mental and some is physical. By *mental*, we mean planning the project and proceeding in an efficient manner so that you are not pushing in the wrong direction. This has to do with performing in the right order the many small tasks that make up a complete job so that one completed phase of the work does not block another. By *physical*, we mean handling tools and materials in a skillful and precise manner so that the correct amount of force is brought to bear in ways that produce good results.

## From Service to Finished Installation

Some examples will clarify these points. Building an electrical service is not too difficult, but if the job is not planned carefully, there is the potential for error and costly rework. We have previously stressed the need to consult with the utility representative prior to starting an installation. The utility will want the meter to be located where it is not exposed to damage (meters are very expensive), where it can be read easily (even if it is a smart meter), and if the service is aerial, where the service drop can connect to the power pole without an obstruction, as shown in Figure 7-1. This is an example of where prior planning makes sense.

Moreover, you have to make sure that your service-entrance conductors can get into the building. If it is a back-to-back installation, make sure that the loca-

FIGURE 7-1    When designing an aerial service, a clear shot at the transformer is needed.

tion of the entrance panel will work on the inside of the building. This enclosure, shown in Figure 7-2, has to be located so that the required dedicated and working spaces are provided; the door can open at least 90 degrees; there is no Code violation such as placing the entrance panel within a bathroom, clothes closet, or on a stairway; there is no potential conflict with water piping or duct work; and preferably it is not necessary to move a framing member.

## Possibilities for Error

This is an example of planning for an electrical installation, and it is similar to many other situations that arise. For every switch or receptacle that you place, there is the possibility for error. Often there is a good, better, and best solution. Even minor decisions can affect the quality of the job and the efficiency of the workflow favorably or adversely.

**FIGURE 7-2**   When designing a service, the location of the entrance panel must be considered.

Appearance is very important. A conduit riser, shown in Figure 7-3, must be plumb to a very tight tolerance or the job will appear shoddy, even if electrically the error is inconsequential. In the case of an underground service lateral where the raceway connects to a 90-degree sweep and then rises vertically to enter a meter socket, if you complete and backfill the trench before mounting the meter socket, more precise positioning of the vertical riser will be possible. When doing raceway work, shown in Figure 7-4, this is a general principle for the placement of surface-mounted receptacles, switch boxes, light fixtures, and so on. Following this procedure will save a lot of work and make for more accurate raceway installations.

As for the physical aspect, you should always be looking for easier and more efficient ways of doing things. The object is to reduce the amount of labor necessary to complete each task without compromising the quality of the installation. Always strive for accuracy. It takes less energy to make a straight and true cut than to make a crooked one.

**Figure 7-3** Underground raceway placement must be coordinated with equipment position if the conduit riser is to be plumb.

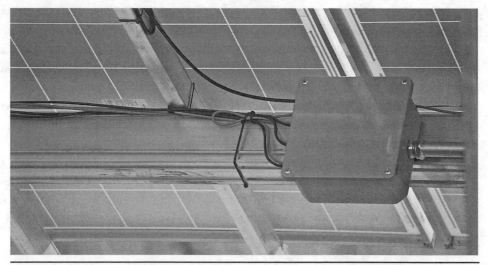

**Figure 7-4** Conduit installations are sometimes used in residential work. For a good appearance, planning is required.

## Learning the Trade

If possible, watch other workers as they ply their trade. If you have the good fortune to work along with a skilled electrician for a period of time, that will be a great help. If you are a school teacher with the summer off, think about it.

If working with or observing a professional is not an option, there are always YouTube videos, shown in Figure 7-5. This is a wonderful way to pick up skills and expertise very quickly. With over 100 hours of video uploaded per minute, there is new material coming online constantly, so check frequently. Type into the search bar, "How to wire an entrance panel," and take it from there. You'll find some misinformation, but the extensive comments section helps to sort it all out. Overall, it's quite lively and worthwhile.

As you gain experience and expertise, even vicariously, as explained earlier, you will become adept at a wide variety of tasks, and you will find that knowledge and skills learned for one type of work can be applied in other areas as well. Use tools to their best advantage. For example, a metal wall box can be grounded by removing at least one of the square cellulose washers from the bolt in the yoke. Instead of slowly unscrewing them from that mile-long bolt, snip them off

**FIGURE 7-5**  YouTube videos are full of valuable information.

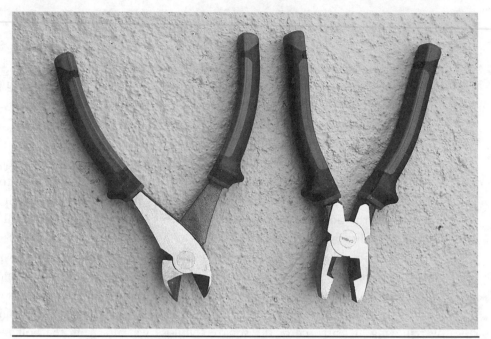

**FIGURE 7-6**    A selection of tools in various sizes facilitates fast, high-quality work.

with your diagonal cutter. And while we are on the subject, removing concentric knockouts from a heavy-duty box such as a meter socket can be laborious in the extreme, but large diagonal cutters (dikes), shown in Figure 7-6, make quick work of this task.

## Some Labor-Saving Tips

Become ambidextrous. Most people favor one hand over the other for intricate tasks that require any degree of dexterity, but you can easily become ambidextrous. All it takes is a little practice. Try hammering nails with your off hand. You will soon get so that you can do it equally well either way. This is a valuable skill when, high on a ladder, you have to reach a long way to both sides.

If you have two pieces to cut and install (it may be wire, raceway, lumber, or something else), always cut the longer piece first. Then, if you make a mistake and the piece proves to be too short, it can be cut down to make the short piece, and you will have a second shot at the long one.

The end of a fish tape will snap off if you try to bend it sharply because it is semihardened. If the end breaks off, heat it with a propane torch to anneal or soften it so that a new hook can be formed.

Often you will want to use electrical tape to temporarily join two or more items. To aid in getting them apart later, spin the end of the tape between your fingers to leave a tail.

When a device bolt breaks off in a wall box, it is a tough situation if the wallboard has been installed. If it is too short to grab with Vise-Grips, drill the broken bolt out, and tap the next size bigger threads. Chuck the tap into your cordless drill, and it will feed in nicely. Use thread-cutting lubricant if you want your tap to last.

If you are going to mount a fluorescent strip fixture on a drywall ceiling, you will have to screw through the enclosure and ceiling material into the framing. The predrilled holes rarely coincide. Screw right through the metal without regard to the predrilled holes, a job for your cordless drill, as shown in Figure 7-7.

Where possible, wire nuts in an enclosure should be positioned with the openings pointed down so that any moisture will drain. If you need to work on a light fixture over a sink, first close the drain so that dropped hardware will not be lost.

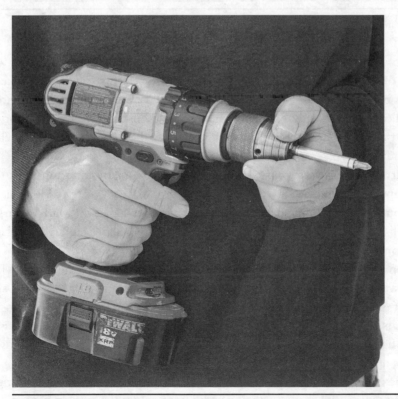

**FIGURE 7-7**   The ubiquitous cordless drill is used in many trades and is indispensible for electrical work.

**FIGURE 7-8**    Bending large conductors for terminations inside an enclosure can be difficult.

Use of 1/0 AWG and larger conductor terminations in a meter socket or entrance panel, shown in Figure 7-8, are difficult. Make use of the hole, with its rounded edges, at the end of the handle of a large adjustable wrench. The moving parts of these tools eventually wear out, so cut off the handles and save them as wire-bending tools.

A partially severed wire buried in a wall will cause an arc-fault breaker to trip out. Temporarily(!) replace it with a standard breaker. A transistor radio tuned to no station with the volume turned high can be moved along the wall. A burst of static will indicate that you have found the fault. If it is a series fault, you will need a load connected at the last receptacle.

When removing Romex, never pull it out from the middle of the roll. It will be full of twists. Instead, take the roll out of the wrapper, and set it up on a short length of pipe supported at both ends so that it can be unreeled. If you are installing Romex through drilled studs, start at the middle and go to the end that is the more difficult route. Then you can unroll enough to get to the other end, allowing for error, and finish the installation.

Romex, shown in Figure 7-9, comes with various color jackets depending on the manufacturer and the lot. If you use different colors for branch circuits, tracing and troubleshooting will be much easier. In addition, Romex handles better

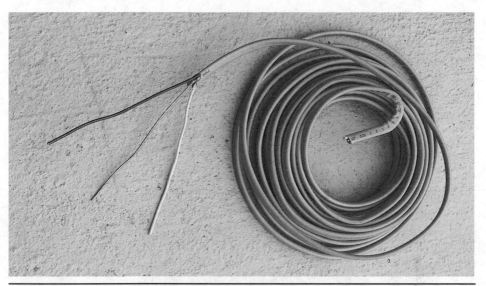

**FIGURE 7-9** Romex cable is available in many colors, and using a selection of them is helpful in tracing branch circuits.

when it is warm. If it is cool, turn up the heat in the room or warm the roll using a small electric heater immediately before installation. A cable ripper is an inexpensive little tool that allows you quickly to slit the outer jacket of Romex without danger of nicking the conductor insulation.

Most wiring can be done without junction boxes. Wherever possible, use daisy-chain rather than spider-web configurations. (The opposite, as we shall see in Chapter 11, is true for communication and data wiring.) Where a junction box is necessary, use a 4 × 4 box, shown in Figure 7-10, rather than an octagonal box. In basements and utility areas, mark the contents of a junction box on the cover.

We won't have much to say in this book about conduit bending because most residential wiring is not in raceways. EMT in small diameters is easy to bend using a conduit bender, as shown in Figure 7-11, but for complex jobs, the routing becomes a daunting challenge—part science and part art.

You need a good eye for what will work and look good, as well as a certain body of knowledge. Conduit bends are shown in Figures 7-12 and 7-13. Complex jobs even call for the use of trigonometric functions and other advanced math. An excellent reference on this topic is the well-known *Benfield Conduit Bending Manual*, 2nd edition (Overland Park, Kansas, EC&M Books, 1993). On the Internet, www.porcupinepress.com also provides lots of useful information on bending conduit.

There are some residential applications where raceways can be used to good effect. We have already mentioned how PVC conduit is perfect for underground

FIGURE 7-10    A 4 × 4 box with an offset bender.

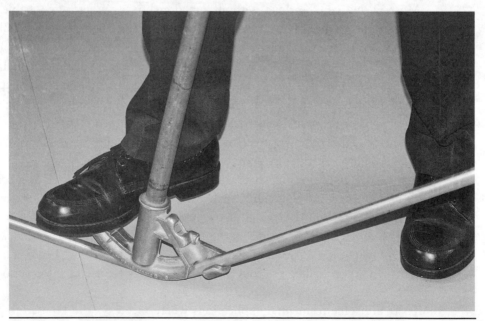

FIGURE 7-11    Bending EMT.

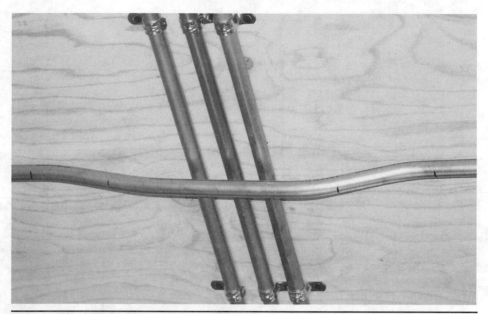

**FIGURE 7-12**    A two-point saddle bend, which is used where two or more pipes must be crossed.

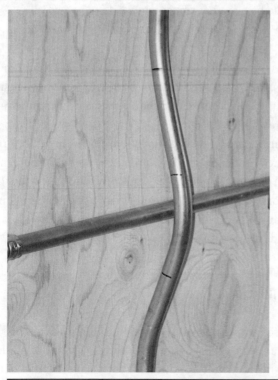

**FIGURE 7-13**    A three-point saddle bend, which is used where a single pipe must be crossed.

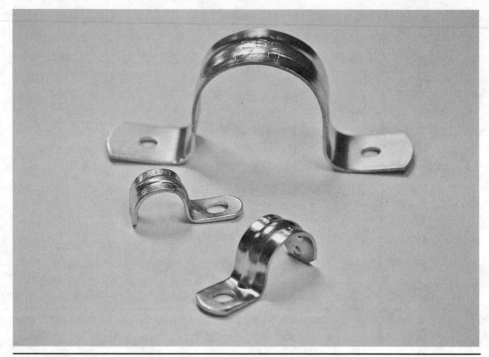

**Figure 7-14**   EMT clamps should be placed at proper intervals.

services and feeders and above-ground outdoor applications where long horizontal runs are avoided. Indoors, PVC also can be used, but EMT is not expensive in small quantities and has a better appearance. It should be secured using the proper hardware, as shown in Figure 7-14.

Anytime you have a wiring drop, such as in a basement or utility room, through the air to a hot-water heater or similar appliance, it should be run in a raceway, preferably EMT. You can transition from Romex to EMT by going through the correct connectors and splicing the conductors using wire nuts inside a 4 × 4 box. Don't forget that the enclosure, at least the cover, must be accessible, although not necessarily readily accessible.

## Comparing Types MC, AC, and EMT

These cable types may be used interchangeably in certain applications, but Type MC cable and Type EMT raceway each have their own design and installation requirements. When used with the right conductors, they have nearly identical uses but quite different working characteristics. For this reason, it is advantageous to use the two types in concert. You have to understand specific construction specifications and permitted uses for each system.

Type MC resembles armored cable (Type AC), but there are some distinct differences. Whereas Type AC has a grounding strip in intimate contact with the inside of the metallic sheath, which thus can serve as the equipment-grounding conductor, Type MC has no such strip. Therefore, the outer sheath of Type MC cable does not qualify as an equipment-grounding conductor, but Type MC cable contains an insulated grounding conductor that, when terminated properly at both ends, makes a reliable return path for fault current, facilitating overcurrent protection. The metal jacket supplements the equipment-grounding conductor. It is solidly connected to grounded metal enclosures at both ends so that the raceway and equipment cannot become energized without causing the overcurrent device to trip out.

EMT is an unthreaded thin-wall raceway that is seen commonly in industrial and commercial locations. The raceway is typically made of steel with a smooth galvanized finish or aluminum. With appropriate fittings terminated properly, it can serve as an equipment-grounding conductor in most applications. However, most electricians pull a green wire for everything.

Then, if the raceway were to pull apart, ground continuity for the circuit would be maintained. The sections of raceway on either side still would be grounded from opposite ends of the run. Also, like the metal sheath of Type MC cable, the raceway provides supplementary and redundant grounding where required and greatly lowers the overall impedance of the equipment-grounding conductor.

Type MC cable and EMT have other traits as well, most notably in uses permitted/not permitted, as noted in the *National Electrical Code* (*NEC*). A simple EMT run is quite easy to install, but a complex job can be a big challenge. Multiple runs coming out of a panel have to be in the right order so that they can peel off as required to go to their final destinations without having to cross. The runs have to be straight and uniformly spaced or they will look unsightly and unprofessional. To maintain uniform spacing, a plywood template based on the raceway positions coming out of the box is helpful.

CHAPTER **8**

# Lighting Fundamentals: Design and Installation

In new residential construction and remodeling, it is said that premium upscale lighting is the least expensive way to enhance the value of a building. An example is shown in Figure 8-1. Consider what the addition of a few high-end outdoor fixtures can do for an otherwise nondescript building! Then there is the whole concept of low-voltage landscape lighting.

The object in residential lighting is to create designs that will be comfortable, aesthetically pleasing, and energy efficient. A great many types of fixtures are available from electrical distributors or in "big box" stores, as shown in Figure 8-2. As opposed to large commercial or industrial work, it is not necessary to have a specialized lighting engineer on the job. If you look over some existing installations, perhaps in the homes of neighbors, and make notes on what seems to work and what does not, you will be off to a good start.

There are some very basic principles that apply to the lighting in different areas. In Chapter 3, we discussed three- and four-way switches in some detail. Be certain that you are placing these switches correctly. Everything should be optimized for the end user. In the interest of energy efficiency, the easiest way to cut costs when it comes to lighting is to facilitate turning off the fixtures when the light is not needed or perhaps dimming them.

## Lighting Control

Besides manual switching, lights may be controlled in a variety of ways, and this is particularly applicable to outdoor lighting, although there are distinct advan-

**FIGURE 8-1**    An outdoor light fixture provides illumination and enhances the value of a home.

**FIGURE 8-2**    Attractive light fixtures are available at modest cost.

tages in optimizing the indoor part of the installation as well. A simple device is the photovoltaic (PV) sensor. It is frequently built into outdoor fixtures, or it may be deployed remotely. When the ambient light declines below a certain level, the solid-state switch conducts so that the fixture lights up. Like a thermostat that controls a furnace or a pressure switch for a water pump or air compressor, there is a certain differential between cut-in and cut-out to prevent short cycling. Often a PV sensor has a sliding sheet-metal shield that may be moved to partially block the ambient light for the purpose of adjusting the on-off times. An inexpensive PV sensor works well for a while, but eventually it fails by shorting out so that the light remains on during daylight hours. Generic replacements are available and attach by means of a locknut at a suitable knockout. Electrically, it is a simple two-wire device that connects in series with the fixture's hot terminal.

Another common and very simple arrangement is a timer. An inexpensive unit plugs into a receptacle and may power a lamp through its cord. (Beware of using an extension cord as a form of permanent wiring.) There is also a larger, higher-amperage timer in a metal enclosure that is intended to control 120- or 240-volt loads such as a hot-water heater. It is useful for lighting as well.

Still another variety of lighting control is accomplished by means of a motion sensor that is built into the fixture. It can be a little difficult to calibrate so that it does what it is supposed to do. It's just a question of tweaking the on-board sensitivity, range, and duration controls and adjusting the motion detector. Because it also has a PV sensor that prevents it from lighting during daylight hours, you may have to return to the fixture more than once. Occupancy sensors save energy by turning off interior fixtures if there is no activity in the room for a prescribed period of time.

There are a number of other means for controlling lights. Door switches at one time were widely used to control closet lights, and in commercial and industrial facilities, battery-powered emergency lights come on instantly when the main power fails, bridging the interval before the backup source comes online.

Another energy-saving strategy is to use the type and size lighting that is appropriate for the setting. When energy was cheap, incandescent bulbs with electrically heated filaments produced abundant light that, at 10 cents per kilowatt-hour, made sense. The next generation was fluorescent light, specifically low-pressure mercury vapor in long tubes.

To get an overview of what is involved in residential lighting, here are some definitions used by professionals:

- **Astronomical time switch.** A lighting control device that switches lights on at dusk and off at dawn irrespective of the actual clock time.

- **Chandelier.** A high-end ceiling-mounted or suspended decorative light fixture, frequently used in dining rooms, that is composed of glass, crystal, or other ornamentation in addition to the lamps.
- **Continuous dimmer switch.** A lighting control that varies lamp brightness gradually from maximum output to off.
- **Stepped dimmer switch.** A lighting control that varies lamp brightness in discrete steps from maximum output to off.
- **Fluorescent.** A low-pressure mercury vapor electric discharge light fixture in which the phosphor coating on the inside surface of the lamp emits visible light when exposed to the ultraviolet (UV) radiation produced by the ionized gas.
- **General lighting.** Also called *ambient lighting* and opposed to task lighting, it provides a uniform level of illumination throughout a large area.
- **High-intensity discharge (HID) lighting.** This includes metal halide and high-pressure sodium lights.
- **Incandescent.** A lamp that contains a metal filament, usually tungsten, that emits copious light when electrically energized.
- **Lamp.** Commonly called a *light bulb*, this generally consists of a glass envelope that contains a gas that can be ionized or a vacuum so that there will be no oxygen that would allow the filament to burn.
- **Light-emitting diode (LED).** A semiconducting device that emits light when forward biased.
- **Lumen.** A measurement of the quantity of light emitted by a light source.
- **Multilevel lighting control.** A device that reduces light in a series of steps as an energy-saving strategy.
- **Pendant.** A light fixture that hangs from the ceiling, often supported by a metal raceway that contains the electrical supply conductors.
- **Sconce.** An attractive wall-mounted light fixture.

The larger the building, the more removed it is from everyday experience, especially in terms of lighting design. We cannot all become lighting design engineers overnight, but as we peruse the literature and become familiar with the terminology (e.g., lumens, color rendition, and the like) and sizing conventions, lighting design begins to make sense. Certain other aspects of the finished product are important. One of these is interior painting. Ceiling and walls should be white—use matte instead of high-gloss paint to minimize glare. The floor should be light colored and low gloss. These seemingly minor building decisions can greatly augment good lighting choices to create a more comfortable home.

In a remodeling job, if the old fluorescent fixture has T-12 or T-10 bulbs, change them over to the more efficient T-8s. This involves changing the ballast. Ignore the old wiring, and follow the wiring diagram on the new ballast. The sockets are the same and do not need to be changed.

As a fluorescent bulb ages, it makes less light and more heat. It also draws more current, which burns out the more expensive ballast. Therefore, fluorescent bulbs should be changed well before they fail completely. Large facilities have a fixed schedule for bulb replacement. A sign of aging is that the bulbs begin to blacken at the ends. Such bulbs should be changed before this becomes very pronounced.

Of course, a key design objective is to provide comfortable, productive light at minimum cost to the owner. Proper switching and dimming capabilities are important ways to minimize energy use. Other strategies include resisting the impulse to overbuild and using the most efficient lighting available. (While we have not completely entered the LED era, there is no doubt that it will be the wave of the future.)

In approaching a new design, the first step is to ascertain the light level required. A careful survey of contemplated usage is in order.

Lighting control is a whole area of expertise that needs to be incorporated into the lighting design. Occupancy sensors and timers are highly effective alternatives to manual switching, which can be neglected.

In the field of residential lighting, saving energy is not the sole concern. Studies have shown that the health of occupants is influenced by the quality of the light. And, of course, the primary goal must be to create a safe installation.

In designing residential lighting, it is helpful to consider different areas inside (and outside) the building separately because there are great differences in the types

## Heat Rise in Recessed Lighting

Recessed lighting is an attractive enhancement for any residential electrical project. It is an effective way to illuminate living space, but like all electrical installations, care must be taken to avoid creating a hazard. The can, or housing, that is installed above the ceiling finish material is designed to limit heat transfer to adjacent combustible materials. Recessed lighting cans are available in two variations. An insulation-contact (IC) model may be installed with zero clearance to the fiberglass batts. If the can is not marked, 2 inches of clearance to insulation should be maintained. A recessed light can contains a thermal overload device that interrupts power to the lamp when the temperature reaches a predetermined level. When the light fixture cools, the lamp will go back on. To prevent such cycling, observe the clearance and maximum lamp rating.

and amounts of illumination that are desirable. Lighting for a workshop would be not at all appropriate for a bedroom, and certainly, the reverse is true.

## Kitchen Lighting

In a kitchen, significantly more lighting is required for the countertop area than for the floor space. Lighting should be provided as needed for specific tasks. The cooking area should be well lighted with good color rendition. Recessed lighting is the way to go, with under-cabinet lighting that is task specific and controlled by separate switching. Recessed cans should be in the ceiling directly above the edge of the counter so that the user's shadow will not block illumination of the task at hand. Three-way switching is essential at room entries.

## Bathroom Lighting

If there is an occupancy sensor, a second light fixture should be provided that is not controlled by that sensor so that a user, taking a bath, will not be left in darkness. No light fixture should be located over the bath or shower. At least two fixtures should be provided on either side of the sink.

## Lighting in Other Rooms

Install multiple fixtures that are controlled by separate switches as an energy-saving strategy. Branch circuits should not exactly correspond to room division so that if an overcurrent device trips, the room will not be left in total darkness. In hallways, three-way switching is essential. Low-output lighting is sufficient and economical. In a workshop, adequate well-placed, cool-white fluorescent lighting is the best choice. Workbench areas should be lighted and switched separately.

## Living Room, Dining Room, and Bedroom Lighting

Wall sconces, shown in Figure 8-3, with low-wattage lamps provide excellent mood lighting that uses less electricity. Be sure that three- and four-way switching is in place. Track lighting is used to illuminate wall pictures and other items of special interest. A chandelier over the dining room table will be a thing of beauty at moderate cost. Be sure to install a dimmer switch!

FIGURE 8-3   Wall sconces enhance the value of any room.

In a bedroom, consider wall sconces and track lighting, if appropriate. A ceiling fan with lamp and fan motor on separate switches is desirable. It should not be located over the bed.

## Outdoor Lighting

Timer switches located indoors, perhaps adjacent to the entrance panel or in a separate closet, are an effective means to control outdoor lighting and permit better regulation than a PV sensor. For example, outdoor lighting can be set to switch off at 1 a.m. A motion sensor will provide the end user with outdoor illumination where access to a switch is not practical. Any lighting installed in the vicinity of a pool or pond must comply with the *National Electrical Code* (NEC), Article 680.

## Fluorescent Fixture Maintenance

If a fluorescent fixture is not lighting, don't measure the voltage as delivered to the bulb socket. A high-impedance measurement taken by multimeter in voltage mode will be inconclusive, and it is hazardous to handle those thin wires when they are live. If a bulb is out, flickering, or at less than full brightness,

- Check power to the fixture.
- Twist the bulbs to see if they are firmly in the sockets.
- Change the bulbs.
- Check the internal wiring and sockets for loose connections.
- Replace the ballast.

Old bulbs draw more current and overload the ballast, causing it to overheat. Inevitably, the ballast will fail. Because the ballasts are more expensive than bulbs and more time-consuming to replace, fluorescent bulbs should be replaced proactively.

The bulbs used most commonly are T-12, T-10, and T-8. T-12 and T-10 bulbs were commonplace for a long time. They are thicker than T-8 bulbs. The older bulbs are initially more costly, use more electricity per unit of light, have a shorter life, are more fragile, and are more bulky to store and to handle. It is good to change from T-12s or T-10s to T-8s.

External wire, fixtures, and sockets are the same for both bulbs. In doing an upgrade, besides the bulbs, the ballast must be changed. If there is no internal disconnect switch, one should be installed. The ballast steps up the supply voltage to about 700 volts, so it is dangerous to handle live secondary wires. Changing bulbs is not hazardous because the glass only near the center of the tube is touched, but changing the ballast involves working on the secondary wiring.

The best procedure is to install an internal switch. The high-voltage wiring varies for different ballasts, but fortunately, the ballast has printed on it a schematic showing connections to the sockets. Be sure that you have the correct replacement. An application chart printed on the ballast shows the number and types of bulbs that the ballast can supply. The ballast may be designed to power four T-12s or two T-8s. After the old ballast has been removed, mount the new one in its place. The ballasts may look different, but the new one will always mount without much trouble. One end of the ballast slides into a slot, and the other end bolts in place. Guided by the schematic on the ballast, connect the secondaries to the correct sockets. Finally, connect the power-supply wires. They are much heavier because they carry more current than is present on the high-voltage side. Fold the wiring compactly in place so that the cover will go on easily without pinching the wires, put in new bulbs, and check the operation of the fixture.

# CHAPTER 9

# Appliance Failure, Maintenance, and Repair

The principal large electrically powered appliances that are found in the home include

- Electric range
- Refrigerator
- Washing machine
- Clothes dryer

They are all used on a daily basis, and there is an ongoing possibility of malfunction or failure. The fault will be either electrical or mechanical or perhaps a combination of the two, for they may interact, and one may lead to the other. Where there are moving parts, failure becomes more likely, and then the repair can become moderately difficult, but if you approach the task in an orderly and systematic fashion, the prospect for a favorable outcome is excellent.

## Isolate and Repair

If the malfunction is purely electrical, it becomes a matter of isolating the defective component and repairing or replacing it. With a multimeter, as shown in Figure 9-1, and a modest amount of other test equipment in hand, the home crafter-electrician should not hesitate to work on this sort of equipment.

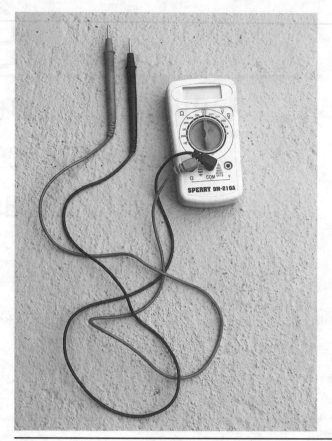

**FIGURE 9-1**   An inexpensive multimeter is adequate for many needs.

The most frequent mistake made by novices is that they think that they will solve the problem in 5 minutes. Because this does not usually happen, they become confused and despondent, the result being either that the appliance does not get fixed or that a professional is called in to make the repair. In this chapter, we'll discuss some approaches to common appliance problems and see how an orderly diagnostic procedure can be applied so that success is achieved.

## An Easy Starting Place

Beginning with the easiest appliance to work on, because it has few moving parts and is easy to get into, we'll consider the electric range, shown in Figure 9-2. It is found in many homes and usually functions for years without incident until one day it fails to perform as expected.

**FIGURE 9-2**   An electric range—long lasting, simple to repair.

An electric range typically consists of four surface burners and, down below, an oven that is accessed through a tilt-down door. The first thing to do is to take a good look at the electric range and determine the extent of the dysfunction. An appliance may be completely dead, or it may power up to some extent but not be totally functional. If it is completely dead, despite the ominous terminology, it may be quite an easy repair.

As always in such matters, check the power supply. Most electric ranges, with the exception of some lightweight studio models, operate on 240 volts. Actually, some portion of the circuitry is designed to operate on 120 volts, so both voltages are required. They are obtained through the familiar three-wire, two-voltage supply from the entrance panel or load center. Through a double-pole breaker, two hot legs (black and red) run to the range. They are accompanied by the grounded neutral (white), which is not fused.

The voltage potential between the two ungrounded phase conductors is 240 volts, and the voltage between either one of them and the neutral is 120 volts. These are nominal voltages, and the measured amounts may differ by 3 percent or more.

The purpose of the 120-volt circuit is to operate the clock, the oven light, and any other incidental loads, as well as to power some electronic circuitry that may be present. Additionally, some models use the 120-volt supply to power some burners at a reduced level in order to obtain a lower heat.

## Shock Hazards in Old Equipment

Besides the three wires, there is a fourth equipment-grounding conductor. At this point, we need to clarify an important matter. Old editions of the *National Electrical Code* (*NEC*) permitted the frame of the range, which is to say the exposed normally non-current-carrying metal parts, to be grounded by means of the neutral conductor. Frequently, 6-6-8 concentric Type SE cable, normally intended as service-entrance cable but permitted for indoor branch circuits, was used to supply power to a range, the outer braided conductive shield doing double duty as the grounded neutral and the equipment ground. Both phase conductors were 6 American Wire Gauge (AWG), and the cable incorporated a reduced 8 AWG neutral. This arrangement has not been permitted in recent Code cycles, which call for the standard three-wire plus equipment-grounding conductor setup. The dilemma arises when an old range that is on a three-wire branch circuit with no equipment ground is replaced by a newer four-wire range.

A range may be hardwired or cord-and-plug connected. A new range normally comes with a four-wire cord and plug (*pigtail*) that is to be field wired to the appliance.

## The Wrong Way

Obviously, the best solution is to run a new four-wire branch circuit with all four wires properly terminated at both ends. What if this is not possible? Well, it is always possible, even if you have to install Wiremold raceway and/or drill through some concrete. If the three-wire cable is to be retained, let someone else do it. In installing the four-wire cord, especially on an older range, be certain that the neutral is not connected via a jumper to the frame, which would violate the principle that the grounded neutral and the equipment-grounding conductor are not to be reunited after leaving the service enclosure. Check it with your ohmmeter. Under no circumstances should the equipment-grounding conductor be floated out to a nearby radiator or isolated ground rod, nor should the equipment ground be picked up from a nearby receptacle.

Returning to our troubleshooting project, if the range appears dead, check the electrical supply at the receptacle or nearest upstream termination, at the range terminals, and, if necessary, at the breaker output terminals in the entrance panel or load center. Sometimes one leg is out because of a faulty breaker connection.

If both legs are not present at the range input terminals, it is a simple house-wiring problem, and you should have the range operational soon. If both legs *are* present at the range terminals but the appliance is not powering up, look for a burned wire or other visual indication that the power flow is blocked.

## Finding the Schematic

If there is no visual indication of an obvious fault, the next step will be to consult the schematic. It may be pasted to the sheet metal at the back of the unit, in a plastic pouch attached to the heat shield, or near the broiler door. If you cannot find the schematic, it should be possible to download it from the manufacturer's website, along with a parts list and service documentation.

A frequent offender is a switch. Many components are actually switches, electrically. With the range powered down, check the components that carry the main power. Then check each control circuit.

If one of the burners is not working, remove it and check it with an ohmmeter, as shown in Figure 9-3. A stovetop burner often plugs into a pair of sockets. Tip it up to clear the base, and pull it straight out. A burner that is faulty may or may not appear burned, but the ohmmeter test is definitive. The sockets where the burner plugs in can be disassembled to check the inner contacts. There is usually a short lead attached, which, in turn, may be burned or broken.

If a wire must be replaced, be sure to use wire with the correct ampacity with high-temperature insulation. The rating, in degrees Celsius, should be on the schematic.

The oven burners are usually bolted in place and best come out with a nut driver. They also can be tested with an ohmmeter. A bad burner will test open.

Generally, the stovetop burners have different heats, whereas the oven burners have a single heat and fluctuate on and off (based on a thermostat) to maintain the correct temperature. Most range problems are easy to fix because the circuits are simple and access is friendly to the user.

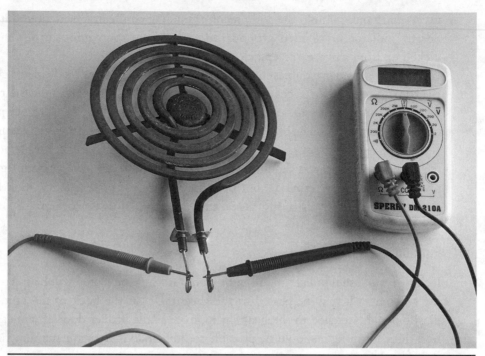

**FIGURE 9-3**    Checking a burner. It should read low ohms.

## Refrigerator

We know how electricity creates heat. Current flows through a heating element, and heat is dissipated according to the expression $I^2R$ (current squared times resistance). But how can electrical energy make the temperature drop?

Consider an airtight cylinder containing a gas that is at the same temperature as the gas outside the cylinder. This gas is made up of molecules that are in constant motion. They bounce off one another like billiard balls, also rebounding off the cylinder walls. The molecular motion is related to the heat energy of the gas. At a higher temperature, the molecular motion accelerates, and greater pressure is exerted on the cylinder's inside wall.

### Heat and Pressure Are Related

If one of the end walls is pushed inward like a moving piston by an outside force, the inside volume of the cylinder decreases. Then the motion of the molecules increases. The amount of heat energy, measured in calories, stays the same, but the temperature, measured in degrees, rises. In the refrigeration process, the gas is compressed in a continuous process, changing phase so that it becomes a liquid and flowing through the circuit. Subsequently, the compressed refrigerant cools, approaching the ambient temperature outside the cylinder.

The cooling process is enhanced by forcing the refrigerant through a finned piping network that resembles an automotive radiator. Ambient air is blown through the core to reduce the excess heat. The stage where this cooling takes place is known as the *condenser*. The cooled refrigerant is still at a high pressure, but the temperature is close to that of the surrounding air. The high-pressure room-temperature refrigerant is allowed to expand, becoming very cold. The caloric heat energy of the refrigerant remains the same, but the temperature drops because the refrigerant is spread through a larger volume. Also, it returns to a gaseous state.

## The Diffuser Valve Makes It Work

The compressed refrigerant does not remain at the same pressure and temperature throughout the entire circuit until returning to the compressor, thanks to the diffuser valve.

This device is placed in-line and marks the division between the high- and low-pressure sides of the refrigeration circuit. It is a pinhole that limits the rate of flow and hence pressure. Downstream from the diffuser valve, the refrigerant pressure is lower. As it expands, it changes back to a gas and becomes much colder.

The refrigerant is piped to a second heat exchanger that also resembles an automotive radiator. It also has a motorized fan that blows air across the metal fins that are bonded to the pipe network. This second heat exchanger is called the *evaporator* because it is here that the process of changing the refrigerant from liquid back to gas is completed. Air blows across the fins and is cooled, maintaining the low temperature inside the box. A thermostat, mounted on the wall inside the box, controls the compressor, causing it to turn on and off at predetermined high and low temperatures. There is a built-in differential to prevent rapid cycling.

The foregoing is a general description of the basic refrigeration process, and it applies to the household refrigerator as well as all kinds of larger refrigeration and air-conditioning systems. Breaking open the refrigerant system to add refrigerant is a job for a licensed technician. This licensing is administered on a federal level in the United States by the Environmental Protection Agency (EPA). Special tools and methods are required to ensure that refrigerant is not released to further damage our fragile ozone layer. Other repairs, where the refrigerant circuit is not opened, can be performed by the home crafter-electrician using simple tools and diagnostic methods.

Household electric refrigerators operate on the basic refrigeration principles. Because there is some variation in the details, it is necessary to consult the nameplate. Find the make and model, and then from the manufacturer's website, download the schematic and service information.

The nameplate is usually found on the back of the refrigerator near the top. Alternative locations include at the top outside edge of the freezer door, inside the food compartment, on the inside of the vegetable crisper or meat drawer, and behind the kick plate. If the nameplate is worn or corroded so that the print is illegible, lightly sand it, and shine a trouble light across it from the side.

## Refrigerator Power Supply

A 15-ampere circuit at 120 volts is sufficient to power most household refrigerators, although they are sometimes put on 20-ampere circuits, and there is nothing wrong with that. Ground-fault circuit-interrupter (GFCI) protection is generally considered incompatible with refrigeration equipment and will nuisance trip at times, possibly resulting in food spoilage. The windings in the hermetic motor-compressor are submerged in refrigerant. A small amount of moisture contamination will create an acidic mix, etching through the insulating coating on the windings. A current leakage to ground will trip out the device. GFCI protection in a residential kitchen is required for countertop receptacles only.

Check the power supply first; then check the power cord. If the refrigerator shows no sign of life including lights, a completely dead refrigerator, the power cord may have become damaged in moving the refrigerator, or it may have been pinched. Check the input terminals inside the unit. If there is voltage there, you know that the problem is internal.

Now you may have to explore different areas depending on whether the unit is dead or has partial power, as indicated by the light. If the bulb is out, test it with an ohmmeter. The center spring terminal in the refrigerator lamp socket may have lost its tension. Unplug the refrigerator, and use needle-nose pliers to pull the spring terminal out to where it belongs. If there is corrosion, use a pencil eraser to clean the metal.

## Door Switch Problems

A door switch controls the light bulb. A small spring-loaded rod extends when the door is opened. See if it is stuck. Check the switch with an ohmmeter with the refrigerator powered down.

If you get the bulb to light, either the control circuitry or the compressor is faulted. Measure the voltage at the motor-compressor terminals. If the motor is not running despite voltage at the terminals, it will need to be replaced. These are sealed units, and ordinarily they cannot be repaired. A licensed refrigeration technician is needed to change the motor-compressor because it involves opening the refrigerant circuit. This repair is rarely done on a household refrigerator because it is usually more economical to buy a new unit.

If there is no voltage at the motor-compressor terminals, look for a fault in the control circuitry. Parts for most models are readily available. They may be generic or from the manufacturer. To perform the diagnosis, consult the schematic. Try a search engine or YouTube, and you will likely find tech forums that will answer specific questions or a video that will show a teardown.

## Checking the Icemaker

Some refrigerators have icemakers. There will be an in-line water filter in the incoming water supply, and it may need to be changed.

All icemakers have some sort of bin switch. It controls power to the icemaker. When the bin becomes full of ice cubes, the bin switch prevents the icemaker from running so that ice cubes will not continue to be produced and overflow the bin. If the bin switch is stuck open, there will be no ice.

A thermal bin switch shuts off the icemaker's power when the ice level rises to contact it. Try warming this sensor. It may be necessary to replace it. A mechanical bin switch is attached to a swivel. When the ice level is low, it drops down, and the icemaker is powered up. The rising ice level pushes the sensor upward, opening the switch.

A mechanical bin switch may stick in the off position. Pulling it down may get it going. If the problem recurs, adding weight by attaching a metal object such as a nut to the sensor lever may be the answer.

# Washing Machine

Washers and dryers, shown in Figure 9-4, are set side by side where possible to facilitate loading wet clothes into the dryer. The doors should open outward. A dryer door usually can be reversed.

The household washing machine is more of a high-maintenance item than either of the appliances we have discussed so far. This is largely due to the fact that there are more moving parts that have greater force exerted on them. There is

**Figure 9-4**   Washers and dryers require a little specialized knowledge to service them successfully.

always an electric motor, usually induction (asynchronous), as well as a transmission that allows for agitator and spin cycles. There are two electrically operated solenoid valves connected by flexible hoses to the domestic hot- and cold-water supplies. These hoses have standard hose fittings with rubber washers.

Many users, to economize, do not use the hot water. It may be shut off at the external supply valve or just not selected at the control panel. There is no pump involved for the hot- and cold-water supplies. They are pressurized by the premises water system.

## An Easy Repair

The cold- and hot-water supplies both have strainers at the inlets. If the machine is observed to be filling slowly, the cause is probably a plugged strainer, especially on the cold-water side. Remove the hose, and you can feel the screen just inside the inlet. If you cannot clean it in place, pull it out with needle-nose pliers and reverse blow it out. Be careful not to pinch or puncture it.

After reattaching the hose and pressurizing the line, the joint may be found to leak. Obtain a new rubber washer, identical to the one in a garden hose, and you're good, unless the threads have been stripped because of cross-threading or overtightening.

Sometimes the solenoid valve stops working or develops a crack, causing it to leak. This part is not very expensive, and it is easy to replace. For an economy job, if hot-water operation is not desired, it may be possible to switch the hot and cold solenoid valves. This cannot be done in some newer models that have dual solenoids built into the same body. You may, however, be able to switch the hoses and electrical terminations. Before condemning the solenoid, check the voltage at the terminals to make that sure it is getting the right signal.

The drain, unlike the cold- and hot-water supplies, has a pump driven off a V-belt. The drain line consists of a flexible hose with a J-bend at the far end so that it can be hooked onto a vertical drain line that goes into the sewer line or to a separate gray-water dry well.

If the eject pump is not working, the drum may be emptied prior to servicing by lowering the drain line so that the water runs out by gravity flow into a floor drain or is hauled away in buckets. (When the pump has failed or is not being driven, water will run freely through it.)

If the washing machine is to be subject to freezing temperatures, it will have to be drained because the pump and associated piping do not drain completely at the end of the eject cycle. For a top loader, tip the machine down so that it lies on a soft pad, and remove the hose from the pump. It is held in place by a very strong spring clamp that may be released (with difficulty) using water-pump pliers. There is a specialized hand tool that makes this task easier. Slide the clamp back along the hose, and pull the hose off the pump. It may be necessary to roll the machine a slight amount from side to side to get all the water out of the drain circuit. Be sure to reattach the hose before putting the washer back in service. The pump is usually belt driven. If it doesn't pump, look for a broken or slipping belt.

If the motor shows no sign of life, check for voltage at its terminals. If it makes a humming sound but does not turn, the transmission may be seized or the drum obstructed. Another possibility is a bad-start capacitor, as shown in Figure 9-5.

If there is no voltage at the motor terminals, assuming that the 120-volt supply has been checked out, there is likely a problem in the electronics. Troubleshoot the timer using a schematic if it is available. The timer usually has spade-type electrical connectors and is easy to change, a common repair.

If the motor runs, but agitator and/or spin action is not taking place, check the belts. You may have to unbolt and electrically disconnect the motor to get the new belt(s) in place.

Most washers have door interlocks that interrupt power to the motor when the lid is lifted. This switch may fail, mechanically or electrically. Except tempo-

**FIGURE 9-5**   The electrolytic start capacitor is a frequent offender.

rarily for test purposes, never bypass the lid switch. It is there to prevent injuries, which may be severe.

## Clothes Dryer

Clothes dryers are a little easier than washers to service and repair in part because airflow is more user friendly than water flow and also because the moving parts are fewer in number and less energetic. The power supply is 240 volts, with some 120-volt loading, so it is a three-wire (plus equipment ground) circuit fused at a mere 30 amps. There are electric heat elements and a single motor. It causes the drum to rotate fairly slowly so that the clothes tumble as they dry more efficiently, and it also powers the blower, which is located downstream from the drum. Consequently, the contents of the drum are not pressurized, but instead, they are exposed to slightly less than atmospheric pressure. The air is drawn past the heat elements into the drum through the tumbling clothes, then through the lint filter, and finally along the vent tube to the outside, where any remaining lint is dispersed and carried away by the breeze.

## Clothes Not Drying

The biggest single problem exhibited by a clothes dryer is that the clothes do not dry well. After the timer causes the cycle to end, the clothes are found to still be damp. This is often caused by an obstruction in the airflow. Clean the lint filter, and check the outdoor shroud to see if the coarse screen, which is intended to keep small animals out, has become clogged. In more severe cases, the flexible vent tube has become plugged. It may be necessary to pull the appliance back from the wall, disconnect the vent tube, and inspect it using a flashlight. An effective remedy is to feed a shop vacuum hose in from one or both ends. The problematic thing about a kinked vent tube is the fact that you sometimes cannot see the tube without pulling the machine back from the wall, which makes the kink go away, only to reappear when the dryer is pushed back into place. As an experiment, try operating the machine away from the wall. It may be necessary to shorten the vent tube a little to prevent kinking.

Another cause of poor drying is one or more bad elements. Sometimes you can open the door of a machine that is in the drying cycle and quickly look inside before the light from the hot elements fades.

## Drum Won't Turn

If the motor is running but the drum is not turning, look for a broken or slipping belt. Typically, the belt is very long and goes around the outside of the drum. It is easily replaced.

Most dryer timers consist of a manually operated control that allows the user to select the amount of time in minutes that drying cycle will operate. More advanced models combine temperature and, in some models, humidity controls. To see how all this works together, it is necessary to consult the schematic and manufacturer's documentation. Solid-state electronic machines introduce an additional level of complexity. If there is an alphanumeric readout with an error code, type it along with the make and model (off the nameplate) into a search engine to find tech forums that should provide an answer.

Dryers may exhibit some mechanical symptoms. They usually involve noisy operation that indicates that breakdown is imminent. Drum supports, which vary in number with different models, may wear out, causing an ever-worsening low-frequency rumbling sound.

Similar sounds are emitted if the belt breaks, causing the belt tensioner to contact the motor shaft. Another kind of noise results when a belt tensioner seizes up or the belt tension spring breaks. These are simple mechanical repairs.

On the inside of the drum are removable plastic vanes that cause the clothes to tumble. They are held in place by screws that are accessible from outside the drum. These screws will need to be tightened if a vane becomes loose and starts clattering.

# CHAPTER 10

# Tackling Difficult Projects

A color TV is an enormously complex piece of equipment in a small box. Of necessity, it is tethered to multiple vast broadcast infrastructures. For this reason, it got off to a slow start. Without transmission, there could be no reception, and without reception, transmission would be pointless. Even black-and-white TV didn't become much of a reality until well into the post–World War II technology boom, and color TV did not become universal until after the mid-1960s.

## Color TV Protocols

For color TV, shown in Figure 10-1, to work, there has to be a protocol that is shared by the broadcast and reception equipment. This shared method of operation goes way beyond the audio and video signals.

A lot of it has to do with the scanning procedure. The analog broadcast system in use in North America is known as the *National Television System Committee* (NTSC). The principal parameters are

- Lines: 525
- Fields: 60
- Horizontal frequency: 15.734 kHz
- Vertical frequency: 60 Hz
- Color subcarrier frequency: 3.579545 MHz

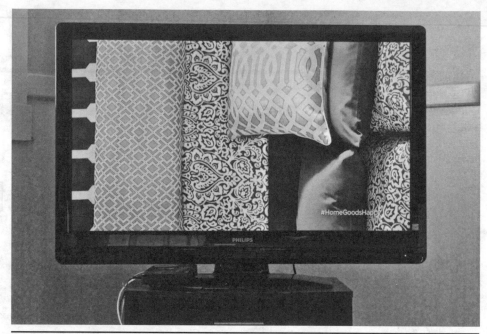

**Figure 10-1**   A flat-screen color TV.

- Video bandwidth: 4.2 MHz
- Sound carrier: 4.5 MHz

For this to work, scanning has to be synchronized between transmitter and receiver. This information must be included as part of the transmission.

What we want to do in this chapter is to introduce some of the basic concepts that underlie color TV troubleshooting and repair. It is way beyond the scope of this book to provide more than an introductory overview. TV technicians know that this work is a very long-range undertaking. As new and more difficult repairs are undertaken, the body of knowledge expands. Technology is also advancing at a rapid pace, so new troubleshooting and repair methods mean that the focus is always shifting. An example is the seismic analog-to-digital upheaval, which has had great consequences for the professional. Despite all of this, we might as well dive in because it is only your old TV that is at stake.

## An Inherent Hazard

That being said, it is a grave mistake to approach this work with a totally relaxed and casual attitude for the simple reason that some of the components are capable of storing lethal voltages long after the TV has been disconnected from the power source.

To begin with the older type of TV set that has a cathode-ray picture tube, to make the picture, a beam of electrons is emitted from a cathode, which is heated by a filament at the rear of this large vacuum tube. The stream of electrons is accelerated toward the screen, where they would make a very bright static spot in the center if it were not for the deflection coils (electrostatic deflection plates in an oscilloscope) that persuade the beam of electrons to scan the screen, side to side and top to bottom. In order to deflect the beam of electrons, high voltages are needed. Inside the TV, a number of different voltages are derived from the 120-volt utility-supplied power. Included are the high deflection voltages.

Among the many components in the TV are electrolytic capacitors. They resemble large cylindrical plastic or metal cans with wire leads or spade terminals. Electrolytic capacitors are also used in motors, either externally or internally, for the start and/or run circuits. Electrolytics are similar to other capacitors, except that they have a very high capacitance and working voltage. The dielectric layer that separates the two plates is not a physical layer of insulation, but instead, it is formed electronically when the voltage is applied. A thin dielectric layer equates to high capacitance, and that is what is needed in power-supply circuits.

Once voltage is applied, the electrolytic capacitors hold the charge for a very long time unless there is a parallel resistance to bleed out the voltage. The lower the resistance, the quicker the voltage will drop. Every capacitor with a parallel resistance has a time constant. What all this means is that when you open a TV cabinet, even if the set is not plugged in, you can really get nailed if you are not very careful.

First, some preliminary precautions are in order. The workbench should be a dry (not oily), clean insulating material such as smooth wood or plastic laminate. You should sit on a nonmetallic stool or stand on a heavy, dry rubber mat or similar nonconductive surface. Receptacles with grounded metal faceplates should not be located along the front edge of the bench but instead along the wall a few inches higher than the bench level where they cannot be accidentally contacted.

Set the TV on the bench, hook up an incoming signal via coaxial cable, and power up the set. Carefully note any symptoms. Is the set completely dead? See if there is a power light that comes on. Even if the screen is completely dark, it is usually possible to sense the presence of voltage by faint light or sound. You may hear a low-level alternating-current (ac) hum coming from the speaker. If there is no sign of life, inspect the power cord from plug to cabinet. Using a tool with an insulated handle, gently tug on and move the cord from side to side where it enters the cabinet.

As a matter of course, use a neon test light to make sure that the chassis is not hot. Most modern TVs have few conductive parts on the outside, but you should be able to pick up a connection to the chassis through the head of a metal screw or the outer shell of the coaxial connector.

Then power down the set by unplugging the cord from the receptacle. Disconnect the coaxial cable signal feed. If, at any point, it is necessary to tip the set forward so that it is resting on the screen, it is permissible to lay it face down on a soft, folded blanket on the bench.

Remove the back from the set. This will disconnect the power cord. TV technicians, of necessity, use a cheater cord to bypass this safety feature so that the set can be run out of its case. They place the TV so that the screen is facing into a large mirror, and in that way, they can work at the back of the set while watching the results. Do not power up the set outside the case unless you have the training and experience to perform this operation safely. Remove any screws that are holding the chassis, and slide it out of the case. Here is where you have to be careful not to contact conductors or terminals that are energized.

Many technicians shunt out the electrolytic capacitors and the high-voltage screen anode (in an analog set) to ground using a screwdriver with an insulated handle or an insulated jumper wire. This is a bad practice because the abrupt current surge can damage the capacitors and other components. It is better to use large low-resistance power resistors with flexible leads connected to insulated alligator clips. Do not remove the chassis from the cabinet unless you are certain that you can find all the high-voltage terminals and safely discharge them. This process has to be repeated after each power-up. The deflection voltages can reach 30 kV, so great care must be taken.

## Preliminary Diagnosis

If you are not an experienced TV service technician, you will be limited as to the types of repairs that you can do. But, by the same token, if the set is completely dead, the possible faults are limited, and troubleshooting, diagnosis, and repair are doable using ordinary electrical tests and repair methods.

First, with the set powered down (unplugged, not just turned off) and all hazardous voltages discharged, visually inspect the area where the power comes into the cabinet. Using an ohmmeter, check each of the power-cord conductors individually, going from the external plug to any power switch, fuse, or other termination. Even without a schematic, you should be able to follow the power flow to the transformer primary. The transformer is a large, heavy laminated steel component, often with cooling fins.

Disconnecting leads as needed to eliminate any parallel current paths, check out the primary and secondary windings of the power transformer with your ohmmeter. The secondaries may be separate windings or multiple taps on a single winding, providing different ac voltages to be rectified and filtered to make direct-current (dc) bias voltages for the solid-state devices and the picture tube.

When disconnected from any possible parallel loads, the individual windings all should read low resistance. Moreover, all leads or terminals should be isolated from the transformer core when none of them is connected to the chassis ground or when the transformer is unbolted and lifted so that it is not grounded.

The bottom line is that if any of the windings are open, the TV won't work. Also, some turns could be shorted internally, altering the output voltage. If primary turns are shorted, it will raise the voltage. If secondary turns are shorted, it will lower the voltage. Because the transformer is not repairable, if it is bad, it will have to be replaced.

Working on a live chassis that is out of the enclosure takes lots of knowledge and nerves of steel, so if you do not have an abundance of both, you will want to limit yourself to some ohmmeter readings with the set powered down and all high voltages bled down to zero. On the Internet, TV service manuals and schematics may be downloaded free of charge, so you will want to obtain this documentation for the TV on which you are working.

In the schematic and a pictorial diagram that is part of the documentation, find the power-supply components, and then locate them on the chassis. The primary parts of the power supply are the diodes that comprise the full-wave rectifier and the electrolytic capacitors that are also part of this network. If any of these parts are not working, the television will not work.

The way to proceed is to visually inspect these components. If a capacitor is swollen, has a distorted shape, or appears charred, it is definitely bad or will fail soon, so it should be replaced. On the other hand, it may look fine but could be defective. The same comments apply to the diodes, except that a faulty diode is less likely to appear bad. For either of these components, check the wire leads, and test them and solder joints for conti-

> **My Books**
>
> Additional information on servicing electronic equipment and the requirements for all kinds of electrical installations may be found in my previous McGraw-Hill books:
>
> - *2011 National Electrical Code®
>   Chapter-by-Chapter* (2012)
> - *Troubleshooting and Repairing
>   Commercial Electrical Equipment*
>   (2013)
> - *The Electrician's Trade Demystified*
>   (2014)

**FIGURE 10-2**    A diode has two leads, and a transistor has three leads.

nuity. Specialized equipment is available for testing capacitors and diodes, but a standard multimeter will give a good go, no-go determination. Figure 10-2 shows a low-power diode and four transistors.

To test a diode, measure the resistance. The lowest range (with the audible continuity beep) works well. Then reverse the leads and measure the resistance going the other way. The multimeter, in the ohms function, applies approximately 3 volts, derived from the interior battery, to the component being tested. A diode has two leads, a cathode and an anode. When a positive voltage is applied to the anode and a negative voltage is applied to the cathode, the diode conducts. It is said to be *forward biased*. When these connections are reversed, the diode does not conduct. It is *reverse biased*. There will be a pronounced difference in these readings if the diode is good. When a diode fails, it often shorts out. Very quickly thereafter, the fault burns itself out so that the bad diode reads open both ways.

The condition of an electrolytic capacitor also can be determined using a multimeter in the ohms mode. It may be necessary to disconnect one terminal so that the device is out of the circuit. Set the ohmmeter to a low-megaohm range for a start. Touch the probes to the two terminals, and note the reading. Then reverse the terminals and compare. You will see that the ohm reading will either remain the same or move in a very stately fashion either upscale or downscale depending on the capacitor's state of charge and the polarity of the probes. The changing readout is very measured and distinctive. It slows as the limit is approached. This is so because the ohmmeter is either charging or discharging the capacitor. This is how a good electrolytic capacitor behaves. Electronic technicians say that the capacitor is *counting*. A bad capacitor will not count. A small-capacitance signal capacitor cannot be checked in this way.

Diodes and capacitors are inexpensive, and replacing them often fixes a power supply and gets the TV back in service. A good many TV malfunctions are as just described. Other problems, such as poor color, may be more difficult to diagnose and repair, and they involve skills and knowledge possessed only by a TV technician. They also involve working on a live chassis, which is not recommended for one who is not fully trained and experienced in the field. However, we may step back and observe an experienced technician at work.

The schematic may have embedded in it small graphics that show waveforms at different points. With a signal generator connected to the tuner or various other appropriate inputs, an oscilloscope can be connected at the points where these waveforms are shown. This requires some knowledge and experience, but for the electronics technician, it is a familiar operation.

The oscilloscope, shown in Figure 10-3, is a voltmeter that depicts the waveform on the screen. Properly calibrated and adjusted and with the correct probe

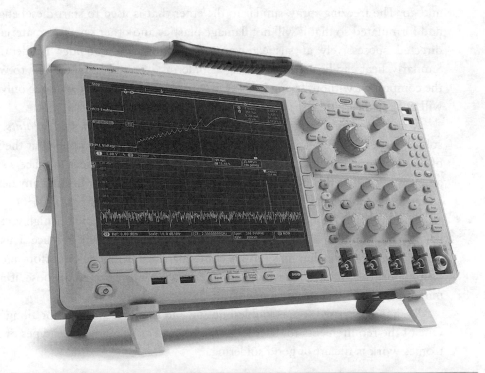

**Figure 10-3**   A digital storage oscilloscope. (*Photo courtesy of Tektronix.*)

chosen, the ground clip is connected to the chassis, and the probe is touched to successive test points indicated on the schematic, looking to see if the waveform appears as shown in the graphics. In this way, the defective stage, circuit, and component can be located. If you are interested in oscilloscopes, take a look at the many Tektronix tutorials, manuals, and data sheets at www.tek.com.

Bias voltages, also indicated in the schematic, can be tested using a standard multimeter. The object, finally, is to identify the defective component(s) so that a repair can be made. All of this, of course, involves working on a live set.

A whole category of faults consists of intermittents. They come and go, raising the uncertainty level and making diagnosis difficult. Most intermittent component failures are caused by changing amounts of heat. As the temperature of a component increases, an invisible crack may open, breaking the current path or causing some other parameter to change. Typically, the set will operate when first turned on and then abruptly exhibit some type of failure.

## Tracking Down Intermittents

Technicians use a product called a *component chiller* to make the symptom come and go. The freezing spray, similar to the ether that is used to start diesel engines and formulated so that it will not damage plastics and other sensitive materials, is directed successively at suspect components until the offender is identified. Similarly, heat can be used by holding a soldering iron just close enough to warm the component without frying it. Of course, this is a diagnostic technique only and will not repair anything.

Many electronic faults are due to what is known as *cold solder joints*. The solder joint, for a variety of reasons, such as insufficient heat applied at the time of manufacture, may fail days, weeks, or months later. The cure is to place a small amount of flux on the joint and remelt the solder without overheating any nearby connected components.

Another effective measure is, with the set powered down and the high voltages discharged, to pull apart each ribbon connector and slide it back in place. This will repolish the contacts that may have become corroded. Ribbon connectors age and become brittle, and if one or more conductors becomes severed or loose at a termination, the set will be adversely affected.

We have discussed some troubleshooting and repair techniques for ailing TVs. One of the requirements for success in this as well as in many other types of electronics work is the art of good soldering.

## Making Good Solder Joints

At one time, all the splices in ordinary house wiring were soldered. The wires to be joined were twisted or looped together and then soldered so as to make a mechanically strong and electrically conductive joint. Then they were wrapped with two types of electrical tape and inserted, along with other similar splices, into grounded junction boxes. When done properly, this type of splice was safe and reliable. But it was time-consuming and required specialized tools and materials. Introduction of the wire nut, an equally safe and reliable solution that was quick and easy, totally eclipsed the old splicing procedure, and that is how it will remain for the foreseeable future.

Despite the fact that soldering no longer plays a role in home wiring, it still figures prominently in electronics fabrication and repair. If you are working inside a TV or other similar electronic equipment found in the home, soldering will be a necessary part of the picture. Fortunately, it is an easy skill to learn, and with a little concentrated study and practice, you can excel.

To make a good solder joint, you need to use the right tools and materials, and you need to have the right techniques. Many metals can be soldered, some more readily than others. Stainless steel and aluminum are impossible to solder without specialized methods and materials. What about brass? It depends on the alloy. Some brass solders easily, and some does not. Fortunately, almost all your soldering will be copper to copper, and that metal solders very nicely.

The best way to learn how to solder is to obtain some scraps of copper and try soldering them together. Then you can put your finished product in a vise and see if you can pull it apart. Also, test the joint with an ohmmeter, stress it severely, and test the resistance again. Similarly, remove one or more printed circuit boards from discarded electronic equipment. Break the circuit board, and then see if you can repair the fracture, soldering appropriate jumpers across broken traces. Remove and resolder components, and test them with your ohmmeter.

Soldering consists of joining the two pieces of copper by heating them sufficiently above the melting point of solder so that when the solder is touched to them, it will melt and bond to both pieces. It is unlike welding in that the two pieces of copper are not melted and allowed to mix. On the other hand, though, the solder does not just sit on the surface of the metal piece to be soldered. It is absorbed a certain distance into both the metals. This is so because copper, like other metals, is highly absorptive. Metal, when it is clean and hot, is like a hungry sponge. Given the right conditions, your solder joints will make themselves.

**FIGURE 10-4**    Wire solder for joining 12 American Wire Gauge (AWG) wire and larger.

Solder generally comes in wire form, as shown in Figure 10-4. For electrical work, it is usually an alloy of tin (Sn) and lead (Pb). Other metals, including silver, may be added, but the usual scenario is an Sn-Pb alloy. In recent years, there has been a move away from lead. For soldering pipes in a potable-water system, lead-free solder is essential. For electrical work, the problem with lead-free solder is that it has a higher melting point. This translates into more difficulty in creating a sound joint and the possibility of component damage due to the higher temperature.

When using solder that contains lead, you definitely do not want to breathe the fumes. Work in a well-ventilated room, and set up a small fan in such a way that it directs the fumes away from you. This is important even when using lead-free solder because, as noted earlier, metals are highly absorptive, and you never know what impurities they may have acquired during manufacture.

The interesting thing about solder alloys is that the melting point is lower than that of either of the constituent metals. The melting point of lead is 621°F. The melting point of tin is 450°F, but the melting point of an alloy consisting of 63 percent tin and 37 percent lead is 361°F, much lower than either of the parent materials. Solder is labeled by using the percent of tin, and the remainder is assumed to be lead unless otherwise stated. For example, Sn 50, used by plumbers to solder copper pipes, is 50 percent tin and 50 percent lead. For electrical work, Sn 60 and Sn 63 are the usual choices.

In order to make a good solder joint, the pieces to be soldered and the soldering-iron tip must be clean, that is, free of dirt, contamination, and an oxide coating. The oxide acts as a thermal barrier, making heat transfer impossible. If there is an oxide coating on your soldering-iron tip, you can bring it up to temperature, but the heat will not transfer to the solder. If you touch a piece of solder wire to a soldering-iron tip that has an oxide coating, the solder will crumble and break into small pieces that fall away, but it will not melt to form a liquid. If you heat a copper wire that has an oxide coating using a clean tip, the solder will melt and then just roll away from the copper wire without bonding. To solder, the tip and the pieces to be soldered must be clean.

To clean the tip, bring it up to temperature, and wipe it on a damp sponge that is kept on the bench for that purpose. This will remove the oxide coating and any dirt or contamination so that the tip will have a shiny, clean appearance. However, at high temperature, the oxide coating will immediately re-form unless steps are taken to prevent this from happening. Immediately after wiping the tip on the sponge, apply a very small amount of flux. When the flux is bubbling, it is at the right temperature. Extended heating will burn the flux, causing it to lose its effectiveness, and the oxide returns.

## How to Tin a Soldering-Iron Tip

Touch a piece of solder to the soldering-iron tip, and as it begins to flow, power down the soldering iron. The solder should flow around the tip, making a protective silvery coating that will not oxidize. This is called *tinning* a soldering tip. It only takes a couple seconds, but it must be done prior to each soldering operation. Also, at the end of the session, the soldering-iron tip should be tinned so that it doesn't corrode during storage and it is ready for the next job.

The pieces to be soldered also need to be cleaned. In storage, copper acquires a dull orange finish, and this is also an oxide coating that will prevent heat transfer and successful soldering. Scrape the wires or terminals gently with a knife without nicking them, or polish them using steel wool. Avoid sandpaper, which can leave sand particles embedded in the metal that will impede conductivity. Apply flux, and as you bring the copper up to soldering temperature, the flux will do the job.

Applying flux is more important than cleaning the work by mechanical means. A freshly cleaned copper surface without flux will reoxidize immediately as it is brought up to temperature, and successful soldering will not be possible.

## What Kind of Flux?

The correct type of flux must be used. For electrical work, definitely do not use the acid-based flux that is used by plumbers to solder copper pipes and is used for automotive radiator work. The correct type of flux is rosin-based flux. It is nonconductive and milder, leaving no corrosive residue that would attack the solder joint or make a conductive bridge that could short out adjacent traces on a circuit board.

Flux comes in a can, and it is applied with a wood or plastic applicator. Cutoffs from installed cable ties are perfect for this use. If you prefer, rosin-core solder containing just the right amount of flux works equally well. If there is any thought that there may be contamination, after it has cooled, the joint can be cleaned with isopropyl alcohol.

Depending on the size wires or circuit board, the correct tip, solder diameter, and technique must be coordinated to achieve consistently successful solder joints. For small work, such as a circuit board, a 15-watt pencil-tip soldering iron is perfect. For larger components, 30 to 50 watts will make a good fit. If you want to solder 12 American Wire Gauge (AWG) wire or larger, a pistol-grip soldering gun, as shown in Figure 10-5, will give good results.

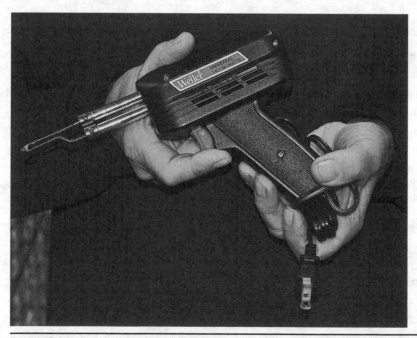

**FIGURE 10-5**    Pistol-grip soldering iron for large work.

For very small printed circuit work, 0.020-inch wire solder is the correct size. As the size of the work increases, use larger-diameter wire solder. For general electronic repair work, 0.031-inch wire is good, and 0.040-inch wire is what you want for soldering 14 AWG and larger copper wires.

To begin with a simple job, let's join two wires together. Strip back the insulation, and clean the copper conductors. Twist or loop them together so as to make a joint that is secure even prior to the soldering. If possible, the prospective solder joint should be suspended in air, not lying on a metal surface that would pull the heat away from the joint. Also, if you were to lay the joint on wood or paper with the soldering tip applied on top, that would be all wrong. Any paper or similar material that scorches or smokes in the vicinity of the solder joint can pollute the operation, making for a poor product.

Another reason for wanting to have the work suspended in free air is so that you can place the soldering iron beneath the joint, because heat travels upward by convection. Also, in this way, the soldering iron does not block your view of the work.

## How to Gauge the Heat

Apply a small amount of flux to the metal to be soldered, and then place the tinned tip of the soldering iron in contact. If the two pieces to be soldered are of unequal mass, apply more heat to the larger piece, proportioning it so that the two pieces come up to temperature simultaneously. On the far (less hot) side of the joint, touch the solder to the work. The solder will serve as a heat gauge, letting you know when the entire joint has reached the melting point of solder. Note the key concept: the solder is not applied to the iron. If it were, you would never know if the actual joint was sufficiently hot to bond well. If the solder is applied to the cold side of the joint and it starts to melt, you will know that there is enough heat everywhere to make a good joint.

Wherever the copper is heated above the melting point of the solder and it has had flux applied to it, the solder will flow. It will readily travel even uphill as long as there is heat, flux, and clean copper. So you don't have to worry about spreading or distributing the solder. Do that by the application of flux and heat. Where the copper is not brought up to temperature, the solder will not flow. If everything is done correctly, the joint will make itself.

Do not overheat the joint. Too much heat will cause the flux to burn, and then parts of the joint will oxidize, and the solder will move away rather than bonding. It should not take more than 2 seconds to make a good solder joint. Get in and get out.

It is very important that the joint is not allowed to move as it cools until it has thoroughly solidified. Any premature relative motion between the two soldered parts will cause fracturing that may not be visible but will make for poor conductivity, arcing, corrosion, mechanical weakness, and similar faults down the road. Don't do anything foolish like spraying water or blowing on the joint to hasten the cooling. This would crystallize the solder, making it brittle and prone to failure. If anything, you would want to retard the cooling for a well-annealed joint.

After the joint has cooled, inspect it with a critical eye. The solder should be a moderately rounded, uniform mass that clings to the work and feathers out at the edges, not too steep and abrupt. It should appear shiny, not dull and foggy, indicating that it has moved during cooling or cooled too fast. It should not have the appearance of wild grapes, indicating that various parts of the solder joint melted and froze individually at different times.

When soldering a component to a circuit board, terminal, or wire, great care must be taken to ensure that the component is not damaged by the heat. Semiconductors, diodes, transistors, integrated circuits (ICs), and especially the ubiquitous complementary metal-oxide semiconductors (CMOSs) are very sensitive to heat, and the problem is that if you destroy one of them in the process of soldering it into a circuit, you will not know immediately because the component will not look any different. Obviously, the shorter the lead, the more acute is the difficulty because the heat has less chance to dissipate before reaching the semiconductor. A good protective measure is to use *heat sinks*. These resemble alligator clips with smooth jaws that fasten onto the lead between the solder joint and the component to intercept and absorb the heat.

The other way to minimize heat damage is to have the soldering iron sized to the job (not too large, not too small) so that it brings the work quickly up to temperature without giving the heat time to travel to the semiconductor. And here again, technique is important. Apply just the right amount of heat, and wrap it up quickly.

So far, we have considered a fairly simple job. Sometimes real-world circumstances conspire to make the soldering task far more difficult. The parts may be very small with short leads, or quarters may be tight so that cleaning is difficult, or it may be impossible to get the soldering iron in at the angle you would desire.

## Printed Circuit Board Repair

Printed circuit board repair presents unique challenges, and great ingenuity is sometimes needed to succeed. Once a defective component has been identified, it must be removed and replaced. As a quick and dirty repair, some technicians, if a component is electrically open, leave it in place and simply bridge a good one over it. Of course, this works, but it doesn't take too much time to remove the old piece. You never know if it might be an intermittent and return to haunt you.

Most circuitry is currently deployed in printed circuit boards. This method is less expensive to manufacture than point-to-point wiring on a chassis, and it is durable and reliable. A familiar task for a repair technician is to remove a defective component from a circuit board and replace it with a new one.

Most circuit boards are of the through-hole variety. The components are mounted on one side of the board, and the leads pass through metal-lined holes that are the electrical contacts, connected to conductive traces printed on the board. On the reverse side, the leads are soldered and the excess trimmed. The solder serves to hold the component in place so that it won't shake loose and it makes a good conductive joint that will hold up for years. The metal contact also serves as a heat sink to prevent damage to the component while it is being soldered.

If you have removed the circuit board from the overall piece of equipment, you will need a third hand to secure it while you hold the solder and the soldering iron. For this purpose, there is a product known as a *helping hand*. It consists of a weighted base fitted with a magnifying glass and alligator clips that can be positioned to hold the printed circuit board or other item to be soldered. Cut short segments of the insulating jacket of Category 5e cable, used for data transmission, and slide these pieces over the alligator-clip jaws so that the printed circuit board is not damaged. On the underside of the board, melt the solder joints, and on the top side, use needle-nose pliers to pull out the old component.

## Clearing Plugged Holes

Very often after an old component has been removed, the holes will be plugged with solder that remained behind. There are two tools that aid in removing the unwanted solder. They are the *solder sucker* and the *solder wick*. The solder sucker takes various forms, as a powered vacuum pump or a hand-operated squeeze bulb, but the idea is the same. You heat the solder to its melting point and quickly suck it out before it has time to refreeze. The solder wick consists of fine-stranded copper wire that is braided and impregnated with rosin flux. Heat the solder to be removed and the solder wick simultaneously, and the wick will draw the solder off the work. The best procedure is to start with the solder sucker and then clean up the remainder with the solder wick. You can also drill out a plugged hole, but this method has the disadvantage that it generates conductive metal filings that may lodge between adjacent traces on the board, shorting them out.

Once the holes in the board have been cleared, shape the leads of the replacement component so that they are parallel and the correct distance apart. Insert them into their holes. If the component has identifying markings, be sure that they face up for future reference. If the component is a diode, make certain that it is polarized correctly.

On the reverse side of the board, bend the leads apart at an angle so that the component stays in place and cannot wiggle. Apply flux. With an appropriately sized tip and solder, make the joints, taking care to apply just the right amount of heat. After the joints have cooled thoroughly without being allowed to move, use your small diagonal cutter to trim the excess leads. Do not make this cut too close to the joint because it could fracture in a way that would not be apparent but could cause trouble for the connection later. If there is any sign of excess flux or foreign material, clean the area with isopropyl alcohol.

This is all there is to replacing a simple two- or three-lead component on a circuit board. To review, keep in mind these potential pitfalls:

- Too much heat will destroy a component.
- If you apply too little heat, a cold joint may result that may work for awhile but will eventually develop increased impedance or fail altogether.
- Too much solder will make a conductive bridge to a nearby trace, shorting out the circuit.
- Any relative motion between the leads and the board during cooling can make a defective joint.

## Removing a Defective Integrated Circuit

A more difficult project is the replacement of an integrated circuit (IC). One common form that it takes is the *dual in-line package* (DIP). This variant typically has 14 or more pins. The problem is that to remove a DIP, every one of the pins has to be brought up to temperature simultaneously. By then, the IC is destroyed by heat. This is not important if the IC is known to be bad, but if the plan is to take it out of the circuit for testing, it must not be overheated. Moreover, the board itself will be damaged by excessive heat.

There are a number of different types of IC extraction tools, and if you are so equipped, that will be the answer. Here's another road to take: in looking over a live TV chassis to find out why it is not working, assuming for now that it does not fall into the completely dead category, you may notice that one of the ICs is hot to the touch. Sample some of the others, and you will get a sense of the normal operating temperature for one of these devices. The high temperature is at once the cause and the symptom, and no matter how you look at it, the IC is finished and will never again function. Therefore, you don't have to worry about damaging it during removal. Using your small diagonal cutter, snip off the pins close to the body of the IC, leaving the stubs that are soldered in place.

An IC socket, many selling for under a dollar, has dual in-line (or some other configuration) pins that are intended to be soldered into a circuit board. Attached to these pins is a socket into which an IC can be inserted and removed any number of times. Solder this IC socket to the cut-off pins from the old IC, and your troubles are ended. Solder a couple pins at a time, allowing the work to cool down in the intervals so that excessive heat doesn't damage the circuit board or the IC.

Trauma, heat, mishandling, and age may affect a circuit board to such an extent that the circuitry is compromised. When you are inspecting a piece of electronic equipment, one of the things to look for is a circuit board that has come loose from its mounts, especially if the set is portable so that it is moved around a lot or if it has been dropped. It is possible that a live terminal or solder joint has grounded out to the chassis or the metal case or frame of some component. The first thing to do is to rebolt the board or whatever it takes and then see if this grounding has overloaded and damaged a trace, wire, or component.

The board itself may have cracked, perhaps just an internal fracture that does not extend to the edges of the board and did not affect any traces or components. If this is the case, drill very small holes at either end of the crack so it does not spread.

Carefully examine any traces that cross the crack. It is likely that one or more is visibly broken. For what it's worth, you can check the trace with an ohmmeter, but even if there is continuity, it is possible that an invisible fracture will worsen with the passage of time, vibration, and changes in temperature. A complete repair will involve rebuilding these traces where there is the possibility that they have been stressed. It will not do just to run some solder into the cracks. This is an incomplete repair that could make the situation worse. Do some point-to-point parallel wiring using 28 or 30 AWG insulated wire between convenient terminals.

Be aware that in high-frequency circuits, any significant change in length or routing could alter the characteristic impedance of the current path. We'll discuss this interesting phenomenon in Chapter 11, but for now, it will suffice to note that length and routing of the conductors should not be altered when the current carried is higher in frequency than an audio signal. Another approach is to rebuild the trace itself. From your electronics supply house, obtain a printed circuit board repair kit. Follow the instructions that come with the kit.

# What About Flat Screens?

So far we have been talking about cathode-ray tube (CRT) TVs as if they were the only kind. In actuality, though, flat-screen technology has largely replaced them. There are two principal types of flat-screen TVs, plasma and liquid-crystal display (LCD). LCD TVs are further subdivided by how they are backlit, by fluorescent or light-emitting diode (LED) lighting. LED flat-screen TVs are sometimes spoken of as if they worked differently from LCD TVs, but the fact is that *LED* only refers to the source of the backlighting. (Plasma flat-screen cells emit light themselves and require no backlighting.)

In a traditional CRT picture tube, a high-intensity electron beam originates at a heated cathode and is accelerated and deflected to scan the phosphor screen. This worked for years and gave an excellent picture, but the flat screen, especially as a computer monitor, is more compact, less intrusive, and far more manageable for most users.

The flat-screen plasma display is made up of an array of pixels placed in grid formation. Each pixel consists of three fluorescent lights—red, green, and blue. Within each fluorescent light is plasma, a gas consisting of electrically charged atoms (ions) and electrons. When not energized, the gas is predominantly uncharged, protons and electrons balancing each other.

## How the Gas Becomes Ionized

It is possible to apply an electric charge by means of external electrodes. The free electrons strike the atoms, which react by losing electrons. The atom now becomes a positive ion. The gas is said to be *ionized*. This is essentially what happens in the familiar fluorescent light bulb. The ballast imparts a high-voltage charge, the gas becomes ionized, and as long as voltage is applied across the length of the tube, the ionized gas glows and emits ultraviolet (UV) radiation, which is converted to visible light when it strikes the phosphor coating on the inside of the glass envelope.

Small cells that are located between two glass plates contain the constituent gases, xenon and neon. On the viewer side are display electrodes that extend horizontally across the viewing area. To the rear are the address electrodes, which are arranged vertically. These two sets of electrodes form a grid.

When the TV's central processing unit (CPU) energizes a horizontal and a vertical electrode, the gas in the cell that is at the intersection becomes ionized and releases UV photons. These particles are invisible to the human eye, but when they strike the phosphor coating on the inside of the cell, visible light is emitted. The ongoing electrical pulses in the electrodes that are connected to the cells correspond to the video information transmitted from far away, and this is how the picture is made.

Each pixel contains three subpixels, each with the appropriate colored phosphor, to impart the color information as intended by the broadcaster. It should be understood that the UV radiation emitted by the cell is not in itself red, green, or blue. The specific color of visible light that will make up the picture is emitted by the phosphor.

The flat-screen LCD TV works in an entirely different way. The pixels may be in one of two states—on or off. This happens by means of liquid crystals that cause polarized light to rotate. In the first place, however, isn't *liquid crystal* a contradiction in terms? Not really. It is a substance that has some of the properties of a solid and some of the properties of a liquid. The atoms can slide around in an unstructured manner, and it can be poured like a liquid, but if electrodes are attached and energized, these atoms will align themselves in a way that resembles a solid crystal. This permits them to polarize light in response to the applied voltage.

You probably know about polarized light. Ordinary light from the sun, a glowing filament, or the flame of a candle consists of a mixture of waves of different frequencies oriented every which way. If you engrave very close parallel opaque lines on a piece of glass, the only light waves that could get through would be those oriented so that the axis representing their amplitude lay parallel to the engraved lines. The lines would be too close to discern. The glass would look just like one of the lenses from a pair of sunglasses. In fact, some (not all) sunglasses have polarized lenses. If you remove both of them from the frame and put them together so that light has to pass through them, the amount of light transmitted will depend on whether the engraved lines are going the same way or are perpendicular. If they are perpendicular, no light will get through. Because of this strange phenomenon, it is possible to make an adjustable light filter from two linearly polarized lenses that are mounted in such a way that one can be rotated with respect to the other.

## LCD Flat Screen

An LCD flat-screen consists of an array of millions of pixels, with subpixels that are red, green, or blue. Each pixel has two polarizing glass filters, one in front and one in back. They are aligned 90 degrees apart with respect to each other. With this arrangement, light will not pass through the pixel. The normal appearance of the pixel, to the viewer, is black.

Unlike the plasma screen, described earlier, the LCD pixels never generate light. The light that makes up the picture comes from the display's backlight, located to the rear of the pixel array. It can be fluorescent or, in a more advanced form, LED light.

How can light pass through the pixel with two perpendicular polarizing filters? It is hard to believe, but the liquid crystal is actually capable of rotating the polarized light while it is in transit within the pixel so that by the time it reaches the second filter, nearer to the viewer, it is aligned so as to pass through it. When voltage is applied to the pixel, the liquid crystal that is between the polarizing filters twists one-quarter turn so that the light that originates at the backlight is able to pass through both polarizing filters. For each pixel, there is a corresponding transistor that is at different times conductive and not conductive so as to rotate the liquid crystal, making the pixel transparent or opaque with respect to the light source behind it.

The question now before us is can we open up one of these flat-screen TVs, plasma or LCD, and attempt to repair it with any expectation of success? Some repairs are entirely feasible. First, obtain the make and model number from the nameplate. Go to the Internet and download the schematic and any service documentation that is available from the manufacturer's website.

Unplug the TV, lay it screen down on a soft blanket that is covering the work surface, and remove the back, generally attached by screws. You will see the principal systems of the receiver. The architecture of a flat-screen TV is simpler and more self-evident than that of the older CRT set, and it is easier to work on.

Like CRT sets, flat-screen models harbor dangerous voltages. They may not rise to the deflection level, but nevertheless, the power-supply capacitors need to be discharged following each power-up cycle. Also, beware of distributed capacitance.

## Other Repair Options

If the TV is completely dead, check the power cord and power supply, as described earlier. You may need to test the power-supply diodes and capacitors using a multimeter in the ohms mode with the set powered down and everything discharged.

Besides the power supply, there is the backlight inverter. If the screen is completely dark but there is sound, there is the high probability that this board has become defective. To test whether the backlight inverter or the backlight itself is bad, you have to either replace the inverter board with a known good one or connect the inverter output to a known good backlight lamp. These parts can be acquired from a discarded unit.

The main board is connected to the video and sound inputs and outputs. You can visually inspect the components for outward signs of damage, but this type of shotgun diagnosis will not often succeed.    There are a number of RCA connectors and removable connectors to power the speakers. The cables that connect boards to each other and to output devices such as speakers and LCDs are likely to be quite fragile and very possibly brittle from age and heat-cycle fatigue, so be careful moving them about.

One repair method that very frequently works is to replace the entire board, but this is expensive and still may not work out. You can pursue a middle course by using the schematic and service documents to do some signal tracing. In this way, if you can narrow the malady down to a single component, you will have made a low-cost repair.

With the limited service resources available to the home crafter-electrician, every TV repair job is not going to end in success, but if you make each project a learning experience, your knowledge and expertise will grow faster than you expect.

## Turning to Computers

These days, almost every home has one or more computers, as shown in Figure 10-6. The price has dropped over the years, but computers still represent a significant investment, so when something goes wrong, there is every reason to see if it can be fixed. Also, there is always the opportunity to purchase or obtain at no cost a fairly advanced machine that has developed a flaw.

A malfunction can be of either a software or a hardware variety, so the first task is to ascertain which of these it is. It is often possible to do the entire repair without going any deeper than the mouse and keyboard. We'll outline some methods for discerning the nature of the malfunction, but first we'll look at the broad topic of preventive maintenance and care.

Laptop computers are very convenient and every bit as functional as their desktop cousins. Because they are more compact, heat dissipation is critical for a long life expectancy.

The biggest single enemy of most electrical equipment is excess heat, and this is particularly important for computers, especially laptops. They pack a lot of electronics into a small volume with a less robust cooling fan and smaller ventilation slots. With nowhere to go, heat accumulates, and the temperature rises. Added to the heat contributed by every component and ambient heat, there is substantial waste heat that is released by the internal battery as it charges and discharges.

**FIGURE 10-6**    A fairly new Dell laptop computer.

Semiconductors, especially ICs, will generate heat in proportion to the number of computations they are required to perform per unit of time. The harder you work the computer, the more heat is generated, and simultaneously, the faster the battery is being discharged. It all adds up.

## Design Strategies

Manufacturers attempt to distribute and place the components so that the temperature rise is not concentrated in one place, but they have other pressing design priorities as well. Components are heat sinked, but neither of these measures removes heat from the cabinet—they just spread it around.

If you want your laptop to last, you have to look for ways to improve heat dissipation, and with a little good management, this can be accomplished. Do not operate a laptop on a bed, blanket, upholstered furniture, or other surface that will impede air circulation. At low cost, you can buy an adjustable laptop stand, shown in Figure 10-7, that will keep the laptop elevated above the tabletop so that air circulates beneath it. Alternatively, you can make a stand from a barbeque grill.

**FIGURE 10-7**   A stand improves air circulation and is highly recommended for long life of the laptop.

This elevating platform will add years to the life of your laptop. Also, when using any computer, locate it in a cool part of the room, away from a heat register and, in summer, near an open window but not in direct sunlight.

When the battery is charging or directly after a work session, do not close the lid. In use, the LCD backlight generates heat, as does the circuitry, keyboard lights, and internal battery, so heat dissipates more easily when the laptop is left open. It should not be closed to put it to sleep. Instead, click on the menu item. Another reason for not always closing the laptop is that there is a ribbon cable or wiring that connects the screen. Opening and closing the lid frequently flexes this wiring, which eventually will wear out, causing a failure.

Do not set drinks or food near a laptop. A spilled liquid is very hazardous to the health of your keyboard and computer circuitry. In the event of a spill, open the case as quickly as possible. Absorb any visible liquid with a dry towel, and situate the exposed circuit boards so they can dry right away. Soda is particularly damaging because when it has dried, a sugar residue remains. Clean between the traces with isopropyl alcohol. Electronic repair shops have special low-temperature ovens used to dry out electronic equipment, and you may be able to improvise something along these lines, but go easy on the heat.

If you go to the beach, leave the laptop at home. Fine sand will find its way into the electronics, and salt-sea air is ruinous, corroding contacts and depositing salt crystals on moving parts. Do not leave your laptop in a closed vehicle in the summer, when temperatures may soar.

Install antivirus software. This helps to prevent software problems that lurk in seemingly innocuous downloads.

## Laptop Battery Maintenance

Proper management is the key to long battery life. Don't fully charge the battery, and don't fully discharge it. If this happens sometimes, it will not be a problem, but you will go more years before buying a new battery if you aim to go up to 80 percent and down to 20 percent of a full charge. Also, the charging cycle generates heat, so it is better to charge the battery in two sessions with a cooling-down interval between them, especially in hot weather.

If you are going to need to operate the laptop under harsh conditions, get a model with a ruggedized case. It will withstand being struck, exposed to moisture, and other adversities.

If a program fails to respond, the wheel spinning continuously, avoid the temptation to cut power to the computer in order to reboot. Instead, force quit the offending program. Every time you power down the computer without a proper shutdown, you put wear on the hard drive. If you do a force quit first, then you can do an orderly menu-driven shutdown. In case the cursor has frozen, use the keyboard shortcut to force quit the topmost application. You should look this up in advance for your particular model and memorize or write it down so that you have it when you need it.

Tobacco smoke is harmful to computers as well as people. If you still haven't quit, get up, go outside. Tobacco smoke is not good for contacts or circuits, and over a period of time, it will discolor plastic parts.

When the work session is done, leave your computer powered up. Just put it to sleep. (In sleep mode, it consumes almost no power.) Powering up a cold machine has been found to put more wear on the hard drive than sleeping it for hours. About once a week, shut it down so that it can get a new reboot and clean up any corruption that may have developed.

Keep the computer clean, and check the ventilation slots for blockage. Use a vacuum cleaner to pull dust out and away from any openings, including optical and other drive slots. Pull dirt that may have accumulated out of all cable connector ports and throughout the keyboard.

## Going Inside

To perform a more complete maintenance, it will be necessary to go inside the box. Go back and review the material on TV servicing earlier in this chapter. The high-voltage safety precautions are relevant to a computer monitor as well as a TV. Specifically, after disconnecting the power cord from the receptacle, discharge all power-supply capacitors, and if it is an old CRT monitor, bleed the charge out of the anode connection when you open the monitor.

Remember that printed circuit boards contain semiconductors, especially of the CMOS variety, that are extremely sensitive to static charge. If you have unknowingly picked up a static charge and you touch one of the terminals or anything that is electrically connected to it, the component will be destroyed. Visually, though, there will be no sign of the damage, so you may have done more harm than good, and you don't even know it. Hold the circuit boards by the edges only. Wear an antistatic bracelet or periodically touch a verified grounded object such as a metal wall plate.

Using a small vacuum cleaner that is made for cleaning electrical equipment, go over the entire inner workings. If there is any noticeable foreign material between the traces on the circuit boards, clean them gently with isopropyl alcohol. Use antistatic wipes to clean other areas.

Unplug each expansion board from the motherboard, clean the contacts with isopropyl alcohol, and slide the board back into place. Where metal surfaces need to be polished, use Scotch Pads. These are the type that are sold in auto parts stores, and they are not impregnated with soap.

If the computer clock has begun to lose time, and the computer is close to five years old, the CMOS clock probably needs to be replaced. In most computers, it is mounted on the motherboard.

## Viruses

If a computer seems unstable, crashes frequently, or is running slower than usual, the problem likely lies in the software domain. Assuming that hardware preventive maintenance, as outlined earlier, has been performed, the next step would be to do a software overhaul. As mentioned earlier, the most important single measure that can be taken is to maintain current antivirus software.

*Computer virus* is a picturesque term borrowed from the field of biology, in which a virus enters a living cell through its membrane, possibly directing it to reproduce in specific ways that are harmful to the host organism. Referring to a computer virus, it is only an analogy, but it is appropriate in that the end result in both cases has the potential to damage the host organism or client machine.

The Mac platform is more resistant than Windows-based machines to infection by viruses in part because Macs are still less numerous than all Windows-based machines combined and in part because the Mac operating system is inherently less vulnerable. There are some very intelligent and highly motivated humans who work hard to devise new viruses that will infect machines on a worldwide scale, the disease being spread by connection to the Internet. Simultaneously, other very capable individuals strive to counter the spread of these infections by devising hardware and software to protect computers from infection. Some antivirus software is available free of charge, and some must be purchased. The antivirus software must match the type of computer and its operating system. It is recommended that the user install and regularly update it.

Another line of defense is to resist the impulse to open any e-mail attachments or links to websites that are not known to be legitimate. Scammers and malicious individuals devise ever more ingenious methods to entice unsuspecting computer users to access sites that are capable of inserting harmful code directly into the user's operating system. A highly effective tool for countering their efforts is the Delete button on your keyboard.

Besides malicious viruses, there are other harmful software threats to your computer, but there are also effective remedies for them. *Malware* includes computer viruses, but it is a larger umbrella term, taking the form of worms, Trojan horses, keyloggers, and the very insidious rogue security software that tricks its victims into buying useless software that is purported to be effective in cleaning harmful malware out of a computer. A free online scanning service may be offered.

Fake timely news articles near the top of a search engine page sometimes take users through one or more redirects to a site indicating that their machine has become infected with malware. A free-trial software claiming to eliminate the malware is offered. Hit the Close tab.

From the foregoing, you can see that the best course of action is to refrain from pursuing any links or offers that you do not know to be genuine. However, it is still possible to have your computer infected surreptitiously.

There are other issues that affect the performance of computers. Some of these are not due to the efforts of malicious individuals but are caused simply by data corruption, where there is no actual hardware malfunction. Every computer has an operating system. It is an integrated set of applications that directs the computer hardware to perform as needed for the user's benefit. Two major operating systems are Mac OS X and Microsoft Windows. As most people on our planet know, the usual architecture for a computer involves various types of memory, either random-access memory (RAM) or read-only memory (ROM) or physical devices for program and data storage, most typically the hard drive.

## Booting Up

The operating system is stored on the hard drive, and as such, it is not available when the computer is first started up. The question, then, is how is it possible for a computer to get anywhere on startup when it has no operating system?

The answer (for Windows-based machines) is the Basic Input/Output System (BIOS). The BIOS is not on the hard drive. It is a separate microchip that is on the motherboard, so it is immediately available on startup. It serves to initialize and test the computer hardware and finally to direct the computer to load the actual operating system from the hard drive. Then that operating system takes over, and the BIOS is no longer in the picture until next time the computer powers up from the off status. Mac architecture is similar, but instead of BIOS, it is called the Extensible Firmware Interface (EFI).

Both BIOS and EFI are examples of *firmware*, which is persistent memory including programming code and data. Firmware exists in nonvolatile memory chips or devices including flash memory, read-only memory (ROM), and erasable-programmable read-only memory (EPROM).

New versions and subversions of operating systems are released as they are developed. The subversions are usually free. The full versions carry a modest price. An upgrade is less expensive than acquiring a new operating system on its own. However, to install a new operating system successfully, previous subversions must be in place. Therefore, it may be necessary to upgrade in stages. Overall, it is a good move to upgrade to a new operating system. Not only does the computer acquire new functionality, but also bugs and corruption that may have accumulated in the old operating system will be eliminated.

## Clearing the Cache

Another good move is to clear the cache in the Internet browser. The procedure varies depending on which browser is being used. Browser programs are offered as free downloads, so in addition to the one that comes with your computer, you may wish to install one or more other browsers. This is a valuable troubleshooting tool because when a site fails to load or interacts poorly, you can try it in another browser to determine whether the glitch is in the website or the browser.

In Safari, the Mac browser, to clear the cache, on the menu bar, click on Safari, and then click on Empty Cache. The cache stores the contents of the Web pages you open, so pages load faster when you return. Other browsers have slightly different methods for clearing the cache, and directions are found by clicking on the Help menu.

Disk cleanup frees up space on the hard drive by allowing the user to manually delete files that will not be needed in the future. It is a good policy to periodically go through your files and move unneeded ones to the trash. Afterwards, empty the trash. The reason for this is that files that are trashed are merely moved to a trash folder, where they can be retrieved later by opening the folder and clicking on the icon. The space on the hard drive does not become available until the trash is emptied. At that point, it still remains on the hard drive but will be overwritten and replaced as the need arises.

As time passes, the hard drive becomes increasingly disorganized and fragmented. Files that are trashed leave gaps here and there throughout the hard drive, and new applications and files grab up prime real estate. The Defrag utility reorganizes the hard drive in the most efficient manner possible.

Windows-based machines include a built-in defrag utility, but current Mac OS X machines do not, so the utility has to be purchased. For today's faster, high-capacity hard drives, there is less need to defrag, and the benefits are less. Also, if you have irreplaceable data that are not backed up, there is the possibility that they will be lost.

## Going Deeper

Let's say that your computer is performing poorly—running slow, experiencing crashes, generally unstable—and you have performed the hardware and software maintenance described earlier to no avail. It is now time to do some serious troubleshooting and repair.

A major indicator is the recent history of the computer. If hardware drivers were recently installed, that can be the problem, and this will sometimes cause the infamous Windows "blue screen of death." Uninstalling a driver may clear the problem.

It is also a fact that programs sometimes conflict. Each program could work fine on its own, but when the two programs are installed on the same hard drive, that's when the problems begin. It is fairly easy to isolate the offending combination by temporarily disabling applications. Rather than doing them one at a time, the most efficient method is to disable half of them for a start and continue in that way until the bad combination is found. This troubleshooting technique is called *divide and conquer*, and it is useful in working with house wiring, data networks, motor repair, and many other areas.

If this doesn't clear the problem, look at the hardware and again the recent history of your machine. If the computer starts to boot up and then crashes partway through the process or it always works fine for a few minutes and then seizes up, there is a strong possibility that there is a temperature problem. Open the case (don't forget the high-voltage precautions), and check the cooling fan. Try to spin the blade by poking through the guard with a pencil. If it won't turn or turns with difficulty, there's your problem. Lots of people believe that a seized motor is "burned out." This couldn't be further from the truth. A seized motor just has a dry bearing, usually the front one. All you need is a little penetrating oil followed up by some heavier machine oil. To do this, if possible, take the fan out of the computer because any excess oil will cause all kinds of electrical problems. When you are done, the motor should spin freely. If it is a valuable computer, you will want to install a new fan because the cost is usually under $25, and the change-out is easy.

If the fan still does not run, the next step is to check its input voltage. Because there may be a little motor vibration, a broken wire is a possibility. Some computer fans are always on from the time the machine is powered up. Others don't receive power until the computer reaches a predetermined temperature. Even where the fan is working, the computer can overheat and crash if there is any sort of airflow blockage, either internal or external, or if it is being operated in high ambient heat.

When a computer starts up, a very distinct tone or chime sounds, and it is a powerful diagnostic tool. It is produced by the BIOS application, and it tells you that no hardware or software problems have been detected. In the PC platform, this is called the *power-on self-test* (POST). The error codes (if any) will vary depending on the type of computer and BIOS microchip. For example, these are the standard original IBM POST error codes:

- One short tone indicates that the system is good.
- Two short tones indicates that the error is as shown on the screen.
- No tone indicates system-board or power-supply failure.
- A continuous tone indicates keyboard, power-supply, or system-board malfunction.
- One long and one short tone indicate system-board failure.
- One long and two short tones indicate a display-adapter malfunction.
- One long and three short tones indicate an enhanced-graphics-adapter malfunction.
- Three long tones indicate keyboard card failure.

To find the meaning of error codes for your particular machine, type the computer make and model into a search engine with query, and you will be directed to a site that should provide an answer.

More advanced software troubleshooting is greatly facilitated by a knowledge of computer programming. An easy starting place is Hyper-Text Machine Language (HTML) coding, used in text files sent over the Internet to direct browsers to display Web pages. This is a very simple form of computer programming. If you are interested in learning about computer programming in general, here is an Internet resource that offers a free in-depth course that will get you started the first day: www.CodeAcademy.com.

## Apple Laptops

In working on computers, a major challenge can be opening the case and later getting it back together without breaking something or marring a finish surface. Some models are vastly more difficult than others. As examples, we'll consider two recent Apple laptop products. The MacBook Pro is relatively user friendly and lends itself to home maintenance and repair, whereas the MacBook Retina is an exacting challenge best consigned to a professional repair shop.

A laptop generally has the same operating system and electronics as a corresponding desktop except for the addition of a battery and charging circuit. However, everything is more compact, and sometimes this makes for more difficult disassembly and parts replacement.

The MacBook Pro 15 inch, shown in Figure 10-8, includes lots of memory and advanced features, but opening the case is simple, and it is not a problem to access battery, trackpad, RAM, hard drive, optical drive, airport card, magsafe board, and fan.

This laptop has been around since 2006, and it has been one of Apple's most successful products. From the outside, it looks like the older PowerBook G4 Apple Laptop, but it has a more advanced hard drive and increased amount of RAM.

As with all sensitive electronic equipment, it is necessary to ensure that you do not fry the semiconductors by exposing them to any static charge that you may have acquired. This is an ongoing endeavor because every time you brush across a plastic part, for example, there is the chance that you are picking up a static charge. There are several ways to guard against damaging semiconductors in this way. This danger is more immediate when the air is dry, such as in a heated room in winter. Humid air continuously bleeds any accumulated electrostatic charge off your body. It may help to keep a hotplate with a pot of steaming water nearby. A more certain protective measure is to wear an antistatic grounding bracelet, although some people worry that it will make for an increased risk of shock if the worker contacts an energized wire or terminal. An effective means of reducing the possibility of electrostatic charge is frequently to touch a grounded surface such

**FIGURE 10-8**    Fifteen-inch MacBook Pro, an Apple classic.

as the metal faceplate of a verified grounded receptacle. Also, handle circuit boards by the edges so that you do not touch high-impedance semiconductor inputs. Soldering irons and other power tools should be kept at ground potential.

When working on a laptop, remove the battery at your earliest convenience so that there will not be a sudden unexpected arc. To remove the battery from the MacBook Pro, invert the laptop and set it on a clean, soft surface. At the bottom of the case, near the middle, are two latches. If you slide them toward the back, the battery will be released. In the space behind the battery, you will see five screws. Three of them, with large heads, secure the memory-bay cover.

Remove the screws. They are not all the same length, so note where each one goes. There is nothing like a digital camera for recording step-by-step disassembly so that things go back correctly. Now you can remove the memory-bay cover. Slide it toward the front of the computer. With the memory-bay cover removed, two RAM slots are revealed. There will be either one or two memory modules depending on what you bought with the laptop and whether anything was subsequently installed as an upgrade. Metal clips hold the memory module(s) in place. The modules can slide out once the metal clips are moved out of the way. The laptop can be upgraded from one to two memory modules, providing a huge 2 GB of RAM, and that makes this simple teardown well worth the effort.

The top case can be removed by taking out 10 Torx screws. If you pull out the top case, you will see the cable that connects it to the main logic board. If the top doesn't want to separate readily, slide in a plastic guitar pick, and gently twist it so as not to chew up the edges. This is a good all-around tool for getting into electronic equipment. After the top case is removed, the hard drive can be taken out. To do this, unplug the hard-drive cable from the main logic board. Two screws on the right side hold the hard-drive bracket. The Bluetooth module and bracket come out easily, and then it is possible to disconnect the main interface cable so that the hard drive is free from the laptop.

## A Final Step

When the new hard drive has been installed (if that was the object of this exercise), it will have to be formatted, and a new operating system will need to be installed. Now, at minimal cost, you have upgraded and given new life to a very valuable laptop that others would have discarded.

Take apart the old hard drive just to see how it works. Can you recognize the type of motor? If you have an inquiring mind, you may want to attempt to rebuild the hard drive and put it back into the laptop to see if it works.

Beyond the hard drive, it is a straight shot to replace the Bluetooth module, Superdrive, Airport Express and antenna cables, keyboard, speaker assembly, Magsafe board, logic board, display, inverter board, and other subsystems.

As you can see from the foregoing, the traditional MacBook Pro is easy to disassemble, and components can be readily replaced. The newer MacBook Pro with Retina Display is a totally different machine. It should be left to professionals because specialized tools and procedures are necessary just to get it apart. Changing the battery, which is cemented in place, is not easily done, and the risk is that you will never get everything back together and working.

Some older laptops are also difficult in the extreme to get into. I once worked for hours on an old Acer laptop before I found a YouTube video that showed how to separate top and bottom cases by inserting a metal pin into an invisible hole behind one of the keyboard buttons. Then the price of the parts precluded fixing the old unit.

When it comes to fixing flat-screen desktops, they are a little more difficult than the old-style computers with separate CRT monitors, but with a little experi-

**FIGURE 10-9**   A recent, very upscale Apple computer.

ence and advance planning, the way should be clear. To get an overview before starting, view a YouTube video on the particular model, and you will be on a sound footing. The difficulty is usually moderate but not severe.

A 24-inch iMac, shown in Figure 10-9, requires a couple of large suction cups with handles to remove the glass panel that encloses the front, and the rest of it is Torx screw work. The hard drive has studs that mount it in place, and they may have to be rearranged to adapt the new hard drive to the existing machine. Keep track of which screws are longer than others. If you try to put a long screw into a short hole, either it won't go or it will damage something inside. A good plan is to use a felt-tip pen to mark out the locations of the longer screws.

## Submersible-Well-Pump Troubleshooting

The home crafter-electrician, like all homeowners, has a drinking-water supply. If it is connected to a municipal water system, the pressure is always there, and it requires little maintenance beyond the initial hookup and paying the water bill.

Others systems are owned and maintained by the end user, and these include a gravity system and suction and submersible-pump systems. The gravity system has no moving parts except for the water itself, which flows from a well whose static level is above the level of the highest faucet in the house. It is the simplest water system, requiring no outside energy to operate, and it is virtually maintenance-free if well constructed. Unfortunately, most building sites do not have the upslope water resource needed.

Another type of water system is built around a suction pump. It is usually located inside the building where the water is used, and as the name implies, it draws the column of water into the house and at its output end pressurizes the premises water system.

The suction pump is used to augment a gravity water supply where the drop is not sufficient to provide acceptable pressure and flow. It is also used to pump water into the house from a well that is located downslope from the house. Generally, the suction-pump system is less expensive to build, but compared with the submersible-pump system, there are some drawbacks.

The suction pump is capable of lifting water to a height of 34 feet above the surface of the water in the well. This is a theoretical limit. Even if a very high-horse-power suction pump were to create a near-perfect vacuum, the water column would not rise higher because the lift is produced by the limited atmospheric pressure exerted at the well. It takes a lot of power to approach this limit, so in actual practice, the maximum vertical lift for a suction pump will be not much over 20 feet.

Another problem with the suction pump is that to start it initially or any time after it loses its prime, it is necessary to remove a plug and to add water manually to prime the pump. (It won't pump air.) Typically, it is necessary to prime the pump repeatedly to get it going. A related problem is that on the suction side, anywhere between the pump and the well, any small pinhole leak will allow air to enter, even if the line is underground. And when that happens, the pump loses its prime. If a grain of sand gets caught in the check valve, the water column may drop back, and the pump will need priming. This has a way of happening when you return home after a vacation.

Another disadvantage of the suction pump is that it requires dedicated space, usually inside the house. And then there is the noise anytime the pump is running. Also, a suction pump has a shorter lifespan than a submersible pump. Because it is always under water, the submersible-pump motor runs much cooler than the suction-pump motor, and it is heat that dooms an electric motor to premature failure.

A suction water pump can consist of separate components that are installed and connected in the field, or alternatively, it is available as a packaged unit with the pressure tank and accessories attached to the pump and motor. Either way, the water and electrical connections are the same. The 120- or 240-volt supply is delivered via a branch circuit with overcurrent protection. The motor can be cord-and-plug connected, but the more common installation is a hardwired connection through a disconnect switch to the pressure switch. This is an automatic switch that responds to water pressure at the pressure-tank manifold.

The pressure switch has two adjustments, best made by means of a nut driver. One adjustment is for the high-pressure cutout, generally set at 55 pounds per square foot (psi), and the other adjustment is for the interval between startup and cutout, generally set at 25 psi. These settings are premade at the factory, and it is usually not necessary to alter them. The tank manifold should be fitted with a pressure gauge so that you can see what is going on.

Wiring is simple. Check the nameplate for the voltage and current rating of the branch circuit. Connect the power supply to the pressure-switch terminals. Because they are flying through the air, the conductors should be in Type FMC raceway or Type MC cable. Connect the pressure-switch load terminals to the motor terminals. This wiring should also not be in Type NM cable.

When first starting up a suction water pump, if you see that it rapidly starts and stops with the pressure-switch points opening and closing once per second or even faster, this is a sure sign that the suction water line has a leak and is taking in air, causing the pump to intermittently lose its prime. Go back along the suction line, inspecting for damage and carefully sealing all joints.

## Advantages of a Submersible Pump

A submersible-water-pump system is preferable from all perspectives except for initial cost. There is no need to prime the pump, shown in Figure 10-10, because as long as it is below the surface of the water in the well, it is always full. And there is not the issue of air entering the line because it is pressurized between the well and the building.

Some people worry that a submersible pump or the wire that supplies it will short out because it is under water. This actually never happens. The pump motor is hermetically sealed, the windings encapsulated in epoxy resin. Heat treating converts this epoxy-resin mix into an impermeable solid so that water can never get to the motor. The whole thing is in a sealed stainless steel can. For this reason, it is impossible ever to rebuild this motor. The other side of the coin, however, is that the motor is incredibly durable and can be expected to run for many years.

Even though the motor cannot be serviced or repaired, it is separable from the pump. They are held together by long bolts, and it is not difficult to separate the motor, which has a splined shaft, from the pump. Then you can attempt to turn

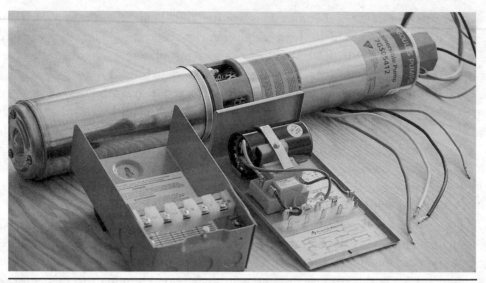

**Figure 10-10**    Submersible pump and control box with cover removed.

the shaft to determine whether the motor is seized. The pump can be easily disassembled and rebuilt by installing a new impeller kit. Sometimes the pump becomes sand bound, preventing it from turning, so the complete repair consists of cleaning it and bolting it back to the motor.

If the motor has seized, shorted, or draws excessive current, a new one can be purchased as a replacement, and this is a much less expensive option than obtaining a complete new pump/motor. There are about a dozen major submersible-pump manufacturers. Most of them use Franklin motors and control boxes. This system is user friendly and easy to work on, with excellent documentation from the manufacturer.

There are two types of installations—two-wire and three-wire. [This nomenclature does not include the equipment-grounding conductor, which recent *National Electrical Code* (NEC) cycles have required to protect the worker in the event that the pump/motor is pulled out of the well and powered up for testing purposes.] If the less costly two-wire option is chosen, the start capacitor and electronics are imbedded in the motor at the bottom of the well. The three-wire installation permits a control box with start capacitor and electronics, as shown in Figure 10-11, to be located in the building for easy access.

In the Franklin control box, these components are contained in the cover, which can be easily pulled off for testing and replacement. Inside the box is a label showing acceptable current and resistance reading, which may be taken at the box or at the wellhead, without pulling the pump/motor. At least 50 percent of the

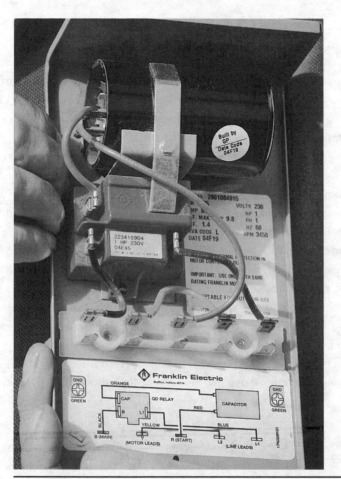

**Figure 10-11**  The start capacitor and associated electronics for a submersible pump are contained in the control-box cover, making for a very easy change-out.

time, a malfunctioning submersible-pump system can be restored simply by replacing the control-box cover.

The output terminals of the control box are wired to the pump motor using pump cable that is color-coded black, red, and yellow (plus a green for the equipment-grounding conductor). Many people believe that the purpose of the three conductors is to provide two hot legs and a neutral for the motor. However, as with most 240-volt motors, the pump motor does not need a neutral.

The red is start, the black is run, and the yellow is common. One of the functions of the control box is to energize the red for a short time to get the motor up to speed, after which the black is switched online. The capacitor is wired into the start circuit. The output lugs at the control box are marked, and the motor has a color-coded pigtail so that you never have a problem knowing how to make the connections.

In the same manner as the suction-pump wiring, the 240-volt supply is brought to the pressure switch, which is a double-pole, in-line device. The load lugs of the pressure switch are wired to the line lugs of the control box. The pump cable begins at the control-box output lugs, labeled "Load." Here again, this wire should be in flexible metallic conduit FMC to a junction box mounted on the wall. Liquid-tight flexible conduit is also used to good effect. Through a polyvinyl chloride (PVC) or metal LB, as shown in Figure 10-12, or knockout in the back of a 4 × 4 box, the pump cable exits the building and goes underground to the well-head. For a first-class installation that is Code compliant, the pump cable should nowhere be visible, including at the wellhead.

A common Code violation is to put the pump cable in black PVC water pipe of the type that is used for underground waterlines. This piece will be seen where it comes out of the ground and attaches to the wellhead, perhaps by means of a hose clamp. If you want to show the world that you know what you are doing, use the gray Underwriters Laboratories (UL)–listed sunlight-resistant PVC conduit with a threaded adapter.

If you are going to pull the pump, remove the cover from the control box. This disconnects the power. Also, lock out the disconnect and shut off the breaker at

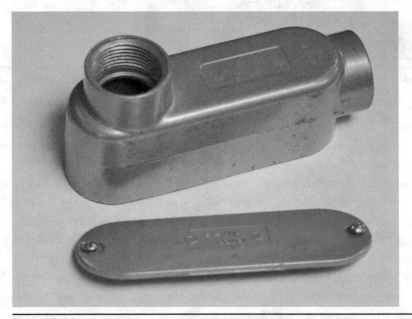

FIGURE 10-12   Type LB fittings are used to make 90-degree bends in a raceway system. The removable cover facilitates pulling the conductors.

the entrance panel. At the wellhead, remove the well cap and disconnect the electrical wires.

A T-handle tool is necessary to remove the pump, but no digging is required. To make the T-handle tool, use black or galvanized 1-inch steel water pipe. You need a piece of pipe, threaded at both ends, that is long enough to reach the pitless adapter, a tee, and two lengths about 8 inches long for handles.

Use a very strong flashlight or trouble light to locate the pitless adapter. It usually will be at the side of the well facing the building, lining up with the burial trench. In cold areas where the frost goes deep, it should be about 5 feet below grade.

Place the threaded end of the T-handle tool into the well casing, threaded end first. Very gently probe around until you contact the pitless adapter. This is a heavy brass fitting that has to be seen to be appreciated.

Screw the T-handle tool into the pitless adapter by turning it clockwise until several threads are engaged. Then, with a sharp upward motion, lift the T-handle so that the inner part of the pitless adapter separates from the outer part, which goes through a hole in the well casing and is attached to the buried water line. The inner part of the pitless adapter is attached to the pipe that goes to the bottom of the well, to which the pump/motor is attached, typically suspended about 8 feet up.

If the drop pipe is steel, a crane will be needed to pull it. If the line is PVC and no longer than 300 feet, it can be pulled by hand, with one or two helpers. Usually the electrical line is taped and/or cable tied to the drop pipe at 5-foot intervals. When pulling the pump, it is essential that the pipe and cable go straight up and out of the well casing without dragging across the sharp steel edge. Any nick in the wire insulation can cause the conductor to ground out, resulting in fast cycling at the control box or worse.

A very frequent cause of submersible-pump-system failure is the chafing of a wire under the well cap. If you leave a moderate-sized loop and tape it lightly, this will not happen. It is in this area that the wires are spliced by means of wire nuts. If they are filled with silicone with the openings pointing down, moisture caused by condensation will not cause the electrical joints to corrode.

The submersible pump for residential use is designed to go in a 6-inch well casing in a drilled well, but it is also suitable for installation in concrete well tiles used in a dug well. But you can't just lay it in at an angle resting on the gravel bottom. To prevent twisting and thrashing about due to reverse torque, the pump/motor must hang suspended a few inches off the bottom. The bearings are designed for vertical operation only, and if the pump/motor sits at an angle, it will have a shortened life.

In constructing a dug well that is going to contain a submersible pump, place the perforated tile through which the waterline is going to enter so that the perforation is at a 45-degree angle to the trench. Using a conduit bender, make an appropriate bend in a piece of 1-inch galvanized steel pipe. This bend will be buried in the ground outside the well tile, and the purpose is so that the pipe cannot turn in the ground, which would allow the drop pipe and pump/motor to move out of plumb.

This steel pipe should terminate near the center of the well, with a 90-degree elbow or tee with a plug attached and pointing down. A vertical-drop pipe should be attached to the elbow and threaded into the pump. The length of this pipe should be such that the pump is suspended a few inches above the bottom surface.

In the horizontal steel pipe, a union should be placed so that the pump can be removed if needed. Bring the electrical line in a PVC sleeve through a separate perforation. Cable-tie it as needed to the water pipe, and leave a loop of slack for a well-organized installation.

# CHAPTER 11

# Data and Communications Networks

D ata and communications networks in the home can range from simple configurations such as a peer-to-peer connection of two computers or two telephones on the same extension to far more elaborate combination Ethernet and wireless structures with many types of machines connected to each other and to the outside world through a modem. If you are getting ready to build a new home or addition, now is the time to design the network so that the cabling will go to the right places, well concealed between points of access. Before describing the mechanics of network cabling, we'll put it in perspective by looking at some of the electronic equipment that is to be connected at both ends of each cable run.

## Satellite Dish Antennas

To start at an upstream input, we'll consider TV and Internet satellite dishes, as shown in Figure 11-1. They are usually mounted on the outside wall or roof of the building or on a pole a short distance away.

It is impossible for electromagnetic radiation to pass through the Earth, so at one time, all broadcasting that went beyond line of sight was restricted to amplitude-modulated (AM) radio signals bounced off an unpredictable iono-sphere or carried by a primitive cable network crisscrossing land and sea. Arthur C. Clark, the great science fiction writer, in 1945 proposed a system of geocentric communications satellites. They would revolve 22,237 miles above the Earth in

**FIGURE 11-1**    This satellite dish is stationary, yet it is always pointed precisely at a geocentric satellite.

circular orbit directly over the equator. To observers on the ground, the satellites would maintain fixed positions in the sky. For this reason, properly aimed satellite dish antennas on Earth could communicate with these unmanned space stations, overcoming the line-of-sight limitation. At present, there are hundreds of these geocentric satellites, permitting Internet access and cable TV in the home.

## Anatomy of the System

In preparing to design a home data network, it is instructive to see how TV and Internet data move from the satellite dishes through the modem and into the home cabling system. The dish is an antenna. Because of its parabolic shape, it has the ability to reflect incoming parallel rays into a single focal point. This process results in passive amplification, the first stage in processing the signal that has been received. But because of the very high microwave frequencies, capacitive and inductive losses would deplete the signal long before it could travel the 30 feet or so to the modem inside the building.

Extremely high frequencies cannot be transmitted by ordinary cabling without great loss. However, still in the form of radiation, they can travel without substantial reduction in strength or distortion along the inside of a waveguide. This is the cylindrical or rectangular tube that you see associated with every satellite dish. The dimensions of this waveguide are based on the frequency of the signal to be conveyed, allowing it to easily pass through the waveguide, bouncing from side to side off the polished inside surfaces. The signal travels the length of the feedhorn to a pickup probe, a small antenna attached to the low-noise block (LNB), as shown in Figure 11-2, which provides further amplification and reduces the carrier to a lower frequency by beating it against a specified frequency generated by a local oscillator.

Because there are semiconductors inside the LNB, a direct-current (dc) power-supply voltage is needed at the dish. Running from the modem inside the building is a length of coaxial cable that carries the power-supply voltage to the dish. This same coaxial cable carries the signal traveling in the opposite direction.

If the purpose of the dish is to provide TV reception, there is no need for transmission, so a single coaxial cable connects it to the receiver inside the building. If it is an Internet access system, two coaxial cables, joined together as a twin cable, run out to the dish. One coaxial cable is for reception, and the other is for transmission.

The home crafter-electrician should become adept at working with coaxial cable, as shown in Figure 11-3. This versatile type of cable gets its name because the outer jacket (a grounded metal sheath that serves as the return conductor), dielectric insulating layer, and inner conductive pin all share the same axis.

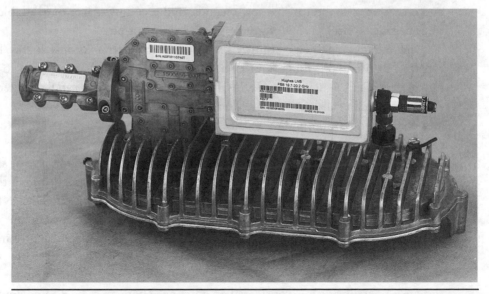

**FIGURE 11-2** The low-noise block performs several functions associated with processing a satellite transmission.

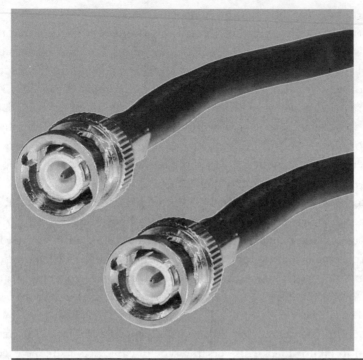

**FIGURE 11-3**    Coaxial input and output for a satellite dish modem.

It is user friendly. With a round profile and just the right stiffness, it is easy to fish through small holes and tight places. Using the right tools, you can cut, strip, and crimp on a connector in a few seconds, and a complete line of available fittings permits multiple types of installations, indoors or out.

The old die-type coax crimper made unreliable terminations. It has been replaced by a compression crimper, shown in Figure 11-4, that is a snap to use, makes watertight joints, and ends the problem of connectors that fall off later on.

If the satellite dish is mounted on the house, secure the coax to the wall, and connect it to a grounding means, as shown in Figure 11-5, adjacent to the entry hole.

You can drive around suburban neighborhoods and get a good sense of satellite installations in all their diversity. A lot of them are not grounded, and of those that are, many are not fully compliant with the *National Electrical Code* (NEC), Chapter 8, "Communications Systems." The key concept is that satellite dish installations are to be grounded and bonded to the building electrical grounding system.

Part II of *NEC* Chapter 8 deals with receiving equipment, including antenna systems. A satellite dish is considered an antenna. There must be a listed antenna discharge unit, which is the grounding means. It can be outside the building or

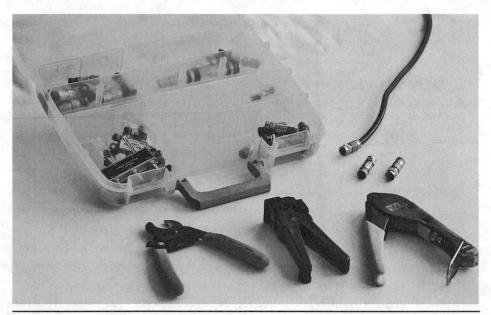

**FIGURE 11-4** Coaxial cable with a connector, tools including a crimper, and an assortment of fittings.

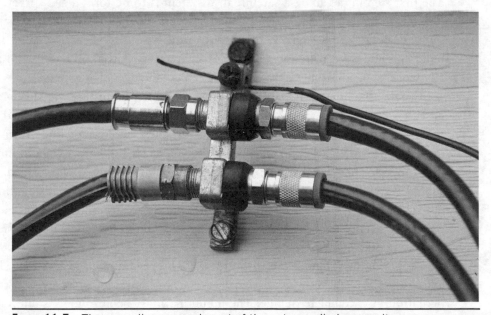

**FIGURE 11-5** The grounding means is part of the antenna discharge unit.

inside as near as possible to the conductors' entry. It is not to be located near combustible material.

## Grounding a Satellite Dish System

The antenna discharge unit has a grounding lug. Connect it to a 6 American Wire Gauge (AWG) bare or insulated, stranded or solid copper conductor connected to the premises grounding system, if it is no farther than 20 feet. If the distance exceeds 20 feet, a separate grounding means, usually a ground rod, is to be provided adjacent to the antenna discharge unit. In this case, the dish grounding means is still to be bonded to the premises electrical grounding system. Use 6 AWG copper, and run it in as straight a line as possible.

If it is a TV satellite dish, the coax is brought into the building and connected to auxiliary equipment provided by the dish manufacturer and then to the receiver, by which is meant not the TV itself but a cable box placed next to the TV and controlled by the user's remote. The receiver is connected to the TV through a short length of coax or color-coded RCA cables and jacks to provide video and stereo sound.

## Getting a Good Signal

The first step in designing a satellite dish system is to perform a site survey. The dish has to be mounted sufficiently high to access the microwave signal from the satellite. Because this signal cannot pass through solid objects, including hills, buildings, and trees, there must be line-of-sight access. A roof or wall mount usually works, but sometimes a pole is needed to get the dish high enough.

Usually, the homeowner subscribes to TV or Internet dish service by signing up either online or at a retail store. Installation may or may not be included free of charge depending on a large number of variables. The vendor often turns the installation over to a nearby satellite dish installer. If the homeowner wants to perform his or her own installation, a certain amount of knowhow is needed. You should be aware at the outset that an Internet access satellite dish installation is more exacting than a TV hookup, and additional steps are required in configuring the connection to the home network. For an experienced technician, the installation could take half a day, less with a good helper.

Mounting and aiming the dish are crucial. If the aiming is off by a slight amount, you might get a good signal at first, but any rain or snow will cause pixilation or a complete outage. The mounting must be absolutely solid; otherwise, a strong wind or the passage of time will throw the dish out of alignment.

Once the dish has been connected to the TV or computer, an on-screen signal-strength meter will appear. A helper, viewing the screen inside and equipped with a cell phone or yelling, can communicate with the worker who is aiming the dish. Another possibility is to take a portable TV or laptop (with a wireless connection to the modem) to where it can be viewed at the dish location. Do not think, however, that you can sweep all over the sky until you hit on the satellite. That would be like trying to hit a small target 22,237 miles away. Furthermore, you would be likely to lock onto some other satellite that would not peak out. The best procedure is to find the correct initial settings for your location, either online or from the vendor, and then use them as a starting point. Rather than using the on-screen signal-strength meter, professional signal meters that connect at the dish are more accurate and easier to use.

An Internet access satellite dish connects to the modem, shown in Figure 11-6, inside the building through two coax lines, one for transmission and one for reception. Unlike the TV satellite dish setup, the Internet access modem connects to the computer using unshielded twisted pair (UTP) cabling, making use of a

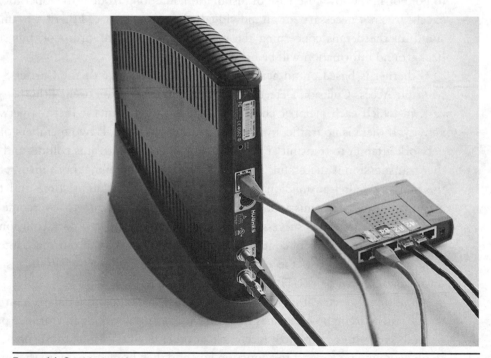

**FIGURE 11-6**  Modem for Internet access satellite dish system. Notice the two coaxial cables, one for receiving and one for transmitting, that come from the dish. A Category 5e cable feeds the modem output to an Ethernet hub, which is necessary because two computers are connected.

communication protocol known as Ethernet. Like coax, UTP is widely used and quite easy to deploy. For the home crafter-electrician who intends to include data networking as a part of the wiring project, a knowledge of both these types of cabling is essential. The good news is that the skills necessary for working with them are easily acquired.

## The Importance of Ethernet

When the modem associated with a satellite dish is connected to one or more computers, or the computers are wired to each other or to printers, the configuration is known as a *local-area network* (LAN). This is generally confined to a single building, but sometimes it spills over into nearby buildings under the same ownership. A larger network is known as a *wide-area network* (WAN), which may be spread throughout a municipal area, extend coast to coast, or cross international boundaries.

When the home crafter-electrician decides to build a LAN, Ethernet is the way to go. For initial low cost, ease of installation, and long trouble-free operation, it excels. It is not necessary for an individual who wants to wire Ethernet to understand all the details concerning the inner workings of the protocol, but some background information will be helpful.

Ethernet is based on an acronym, CSMA/CD, that stands for Carrier Sense Multiple Access Collision Detect. This is a precise description of the Ethernet process, in which each member constantly listens to determine if there is network activity. If there is no traffic, transmission may commence. If two members of the network attempt to transmit at the same time, their data streams collide, and neither transmission is successful. To prevent this from happening, both members of the network halt transmission. They wait different amounts of time and then attempt another transmission. This process enables Ethernet to overcome the principal impediment to successful network operation, which is data collision.

The Ethernet Protocol contains other elements as well. But our main interest is in a different part of the equation, which is the Ethernet medium, usually unshielded twisted pair (UTP), as shown in Figure 11-7.

Traditional electrical work is serial in nature, arranged in a daisy-chain pattern. Branch-circuit power comes from an overcurrent device in an entrance panel or load center. It goes first to a receptacle or other device, then on to the next, and so on. Running power first to a junction box and then branching out to the outlets is permitted by the *NEC*, but this is not the best way to do it.

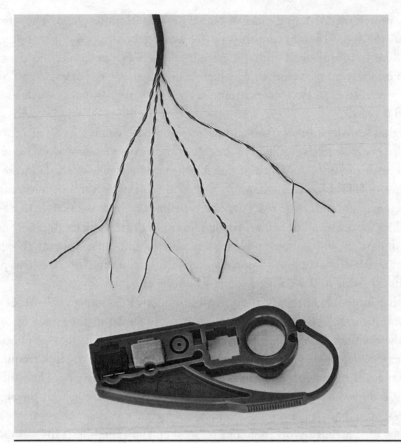

**FIGURE 11-7**    UTP, in this case Category 5e, has four twisted pairs of conductors. The tool makes quick work of precisely stripping the jacket so that there is never a nicked wire.

For Ethernet, the preferred configuration is star topology. More cable is required to wire each branch back to the source, but in data networking, this pattern is better. There are no extra splices, which are undesirable because they disrupt the twisting of paired conductors. And twisting is essential because it counteracts inductive loss.

## How Twisting of Conductors Helps

Twisting of data conductors in UTP is actually the principal strategy that permits high-speed connectivity. At one time, only teletypes and printers were connected. Now, even in the home, we expect to download graphics, music, and large video files, so high-speed connectivity is required.

Transmitter, medium, and receiver comprise a continuous chain. The slowest link limits the speed of data transmission. To achieve high-speed connectivity, we need to find the bottleneck and eliminate it. Traversing the cable, the data are made up of a stream of bits, each of which is in either of two states, on/off, true/false, or one/zero. The two states are represented by two voltages, such as +12 volts and 0 volts.

In digital transmission, the transitions between these voltage levels are abrupt, practically instantaneous, unlike the graceful sine-waves characteristic of analog signals. For this reason, a high twist rate for the conductors is needed, and this is the essence of UTP. Because digital transmission involves high frequencies, capacitive reactance and inductive reactance become limiting factors. Capacitive reactance, which is measured in ohms, becomes less at high frequencies, but because it appears in parallel with source and load, it tends to shunt out the signal, the effect increasing with frequency. Additionally, capacitive coupling of adjacent pairs makes for crosstalk and a reduced signal-to-noise ratio.

Another impediment to signal integrity is inductive reactance. Whenever current flows through a conductor, a magnetic field is set up in the space that surrounds the conductor. Where the current is alternating current (ac) or pulsating direct current (dc), energy is required to move the magnetic field. High frequency, long distance, or both conspire to make the effect more pronounced. The continuous work required to move the magnetic field reduces the strength of the signal. In contrast to capacitive reactance, inductive reactance is a series phenomenon. Both types of reactance work together to weaken the signal, and the combined effects are greater at a higher frequency.

Shielding is one way to lessen the harmful effects of capacitive coupling. One method is to install the cable inside electrical metallic tubing (EMT), which is always grounded. This strategy reduces capacitive coupling to adjacent conductors and results in a better signal-to-noise ratio. Challenges involved in high-frequency data transmission also may be met by scrupulously matching impedances so that maximum power transfer will occur. Harmful reflections take place when there is a mismatch in source and load impedances. Then the digital waveform is distorted, and the signal may become totally unreadable.

Another thing that may have to be watched in large buildings is that maximum distances must not be exceeded. Using Category 5e UTP, the maximum distance is not much over 300 feet.

As current rises or falls in a conductor, as we have noted, the magnetic field moves as well. Then current is made to flow in nearby parallel conductors. The conductors, in effect, are primaries and secondaries of a transformer. This phenomenon is known as *mutual inductance*. A related phenomenon is *self-induc-*

*tance*, which occurs when current flow is induced in the original conductor. This secondary current is opposite in polarity, and it is subtracted from the primary current. The end result is that the conductor exhibits impedance to the flow of current in proportion to the frequency. Like resistance, this impedance is measured in ohms and conforms to Ohm's law, the difference being that it is frequency dependant. Of course, it increases in magnitude with the length of the transmission line.

## Wire Transposition

Nineteenth-century telephone system designers employed balanced circuits to reduce loss and extend transmission range. They were able to achieve long-distance transmission by means of wire transposition. At every third pole or so, the wires would cross and exchange places. The twist rate, about five turns per mile, along with balanced transmission, greatly improved the signal-to-noise ratio.

In the past few years, there has been a vast increase in the speed of data transmission. The first networks were only required to run teletype machines or connect computers and printers, but now everyone wants to download graphics, music, and movies. Very high-speed connectivity is expected, especially in the home.

In the twenty-first century, we see that twisted-pair technology has become dominant for telephone and data-transmission applications, and it is deployed extensively In home networking. The other widely used medium is coaxial cable. It derives its name from the fact that the inner conductor is centered within the outer grounded shield, which is also the signal-return conductor. They share a common axis.

Coax has been and still is used for antenna transmission, instrumentation, surveillance video, and other applications. It was an early Ethernet medium. The only problems with coaxial and other shielded cables are that they are a little pricey and time-consuming to install compared with UTP, such as the widely used Category 5e, which is perfect for many of our telephone and data needs.

### How It Works

UTP depends for its success on balanced signal transmission. In balanced signal transmission, each of the two conductors carries a mirror image of the same signal, identical but 180 degrees out of phase. Digital information is conveyed by the differential amplitude of the two signals, which may vary from zero to some predetermined maximum value. If a pair of these conductors were run past a source of interference, such as a fluorescent ballast, motor, or an ac power line, one of the conductors, the nearer one, would acquire more noise by means of inductive coupling than the other, resulting in an unacceptable signal-to-noise ratio. By twisting the members of each pair, harmful interference is greatly reduced. First one conductor and then the other is closer to the source of noise so that each receives an equal amount. Because the balanced signal is differential, the noise that is in the data line is canceled. Balanced signal transmission by itself is not totally effective in eliminating noise, but it is necessary for the conductor twisting to be effective.

## Precautions

A tighter twist rate and other design improvements in UTP have kept pace with faster data speeds. Now the focus must be on the quality of the installation. UTP is easy to install and connect at terminations, but care is needed to preserve high-speed performance. Keep these points in mind:

- Do not run UTP close to a fluorescent ballast, motor with brush-type commutation, or an ac power line. Maintain a minimum separation of 6 inches, never sharing the same hole in a framing member. Keep UTP away from uninterruptible power supplies, transformers, or any equipment that generates a magnetic field.
- Leave staples and other hardware slightly loose in securing the cable so that the conductors are not pinched.
- The twist rate must not be altered. Avoid kinks, sharp bends, and too much pulling force.
- If cable is laid out on the floor, do not allow visitors to step on it.
- If it is necessary to fish cable through hollow wall and ceiling spaces, leave extra cable so that any damaged ends can be trimmed off before termination.

It is not necessary to be familiar with all the intricacies of Ethernet Standard IEEE 802.3 to successfully build a data network for the home. The protocol is contained within and implemented by network interface cards (NICs) that are required at both ends of the cabling in order to establish the link. (These Ethernet cards are backwards compatible, but the NIC for fiberoptics is mechanically different because of the termination method.)

Optical fiber cable is definitely on the horizon, but for the present, most home networks employ UTP, specifically Category 5e, because it is well able to provide the speed and reliability needed in this setting. If in the future a decision is made to cycle over to optical fiber, the existing UTP, if it is installed in raceways, can be used as a rope to pull the new cable in place, but as mentioned previously, new NICs would be needed.

## Choosing the Medium

Raceway is not an absolute requirement in most environments, but it is highly recommended as an excellent installation method. UTP cable can be fastened to any wood-finish surface. For drywall, it is best to use screw clips fastened through the sheetrock layer into the framing. An alternative is to use plastic shields sized to fit the screws and set into the drywall.

Wiremold works well where a finished appearance is desired. Where possible, conceal the cable behind wall or ceiling finish. The space above a suspended ceiling works well, but the cable is not to be laid across the panels in a way that would block access. For a retrofit, use fish tapes, chains, magnets, and other installation aids.

Though not required, a top-of-the-line installation involves pulling the UTP through EMT. This very secure metal raceway is coupled and terminated using simple setscrew fittings indoors and compression fittings outside. A conduit bender will make gentle sweeps.

EMT connectors go into enclosure knockouts. One of the advantages of a metal raceway such as EMT is that because it is grounded, there is isolation from radio-frequency (RF) radiation created by harmonics as well as complete fire protection.

If a piece of UTP is damaged in handling or doesn't look good for any reason, don't use it in a data network. Do not use previously installed cable. However, these discards are good for telephone extensions between jacks, where there are no high-frequency issues.

Small wire nuts or crimpable bugs are suitable for telephone work, but where UTP must be spliced in a data network, an Ethernet hub (Figure 11-8), switch, or router must be used. To emphasize, all Ethernet runs must be splice-free. Unlike coaxial cable used for video transmission, a splitter will not do. An Ethernet hub or better is needed.

In designing a data network for the home, we must begin by deciding what type of Category 5e to use and how to install it. Category 5e UTP cable is subdivided as follows:

- It is available in solid and stranded versions. Stranded cable is used for short patch cords that will be flexed or bent repeatedly or where there is the possibility of vibration. Otherwise, solid cable is preferred because it

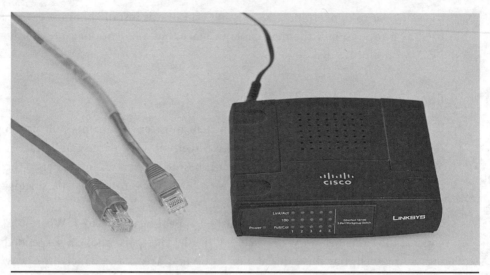

**FIGURE 11-8**    An Ethernet hub is used where a single feed splits into two or more branches. The wire at the top is an ac power supply. UTP with RJ45 modular connectors plug into a varying number of ports.

has better high-speed performance. The two types of cable require different connectors. Both are of the insulation-displacement type. The conductor ends are not stripped. The insulation is pierced by the conductive metal prongs inside the connectors. If the copper is stranded, it is pierced, whereas in solid conductor connections, the metal prongs slide past the conductor and press tightly against it to make the connection. For the most part, you won't be using connectors designed for stranded UTP cable because the short patch cords are generally purchased premade.

- UTP is available for plenum, riser, and general-purpose applications. This has to do with the smoke-generation and fire-propagation characteristics of the UTP cable. (It is also applicable to coax and other types of communication cable.) The amount of current carried in an Ethernet circuit is not capable of igniting combustible material. But the insulation, once ignited by a fire that originates elsewhere, will propagate flame and generate copious amounts of thick, toxic smoke. To control this hazardous situation, it is important to use the correct cable for specified environments, which are ranked according to their sensitivity. Unfortunately, the cables having less hazardous jacket and conductor insulation properties also happen to be more expensive. Plenum-rated category cable is used in any space such as above a suspended ceiling that transports air from one part of a building to another. A riser is required where the cable comprises a vertical run that penetrates from one floor to another. General-purpose cable is for other locations, and limited-use cable is permitted in dwellings only. In all cases, the cable for a more sensitive location can be used in any less sensitive location. Thus plenum-rated UTP cable is permitted in any of the other locations. For small jobs where you are not attempting to cut the cost to an absolute minimum, you may want to keep a single carton of plenum-rated Category 5e cable and use it for everything.

- UTP is also divided into indoor and wet-location cables. The latter is required for underground use in a raceway, which is always considered a wet location because the raceway is presumed to fill with water at times.

## Terminating UTP Cable

At last, we come to the heart of the matter—UTP terminations. Always run the cable first, allowing extra length, before the connectors are attached to the ends. In this way, if the end is mushroomed as it is forced through a drilled hole or around an obstruction, a few inches can be trimmed back to get a good end. Also, extra cable in the form of a loop should be left at each end in case the termination has to be remade in the future.

After the cable is in place, we are ready to attach the RJ45 connectors at both ends. First, slide on a rubber boot. This item is optional, but it improves the quality of the job. It protects the cable where it enters the connector and prevents abrasive material and moisture from degrading the connection.

UTP Category 5e cable contains four twisted pairs within a loose-fitting jacket. The Ethernet Protocol specifies that either two or four pairs are to be used depending on the version. In either of these formats, the pin-out is the same. Any unused conductors are also terminated for future use and for mechanical strength.

Using sharp scissors, trim back the jacket about 2 inches. An inexpensive data cable stripper is quick and accurate, but an electrician's wire stripper using the 10 AWG cutter will work if you take care not to nick the insulation. Note that only the jacket is removed, not the insulation from the individual conductors.

Cut off the rip cord. Untwist the conductors back to the jacket, fan them out to make a flat plane, and arrange them in the proper order depending on which of the two configurations is being used (see below). Straighten the conductors, making them parallel, in the right order, and spaced so that they will go into the connector.

Using sharp scissors or the cutter that is part of some RJ45 crimpers, make the final cut at a right angle across all eight conductors so that they are no longer than ½ inch. If you cut them too long, the Ethernet connection will exhibit crosstalk. If you cut them too short, the signal will be weak, intermittent, or nonexistent.

Making sure that the conductors are straight, properly spaced, parallel, and in the right order, push them into the connector. You should feel them bottom out. The conductors should not buckle, fold over, or exchange places. Using the RJ45 crimper, give a good squeeze on the connector. It is transparent, so you can inspect the final product. Slide the rubber boot into place, and you are done.

As for the order of the conductors, it is necessary to realize that at both ends there are transmit and receive pins. You have to connect the transmit pins at one end to the receive pins at the other end. Sometimes you use straight-through wiring, the terminations being the same at both ends. In other situations, you employ crossover wiring. It depends on what is connected at each end. If you are going from an Ethernet hub or switch to a computer, use straight-through wiring. If you are going from a computer to another computer or from a hub to another hub, use the crossover configuration. If you make a crossover cable, mark both terminations with an X so that there will be no confusion in the future.

## Ethernet Pin-Outs

RJ45 pin-outs for Ethernet connections are mandated according to the Telecommunications Industry Association (TIA) standards. Looking at a connector with the clip to the back and wire opening down, there are eight terminals numbered 1 to 8 starting at the left. For straight-through wiring, used to connect a hub to a computer, T568-B is used at both ends. (T568-A could be used at both ends, but it is customary to use T568-B at both ends for straight-through wiring.) For crossover wiring, used to connect hub to hub or computer to computer, T568-A is used at one end, and T568-B is used at the other end.

T568-A is

- Terminal 1: green/white
- Terminal 2: green
- Terminal 3: orange/white
- Terminal 4: blue
- Terminal 5: blue/white
- Terminal 6: orange
- Terminal 7: brown/white
- Terminal 8: brown

T568-B is

- Terminal 1: orange/white
- Terminal 2: orange

- Terminal 3: green/white
- Terminal 4: blue
- Terminal 5: blue/white
- Terminal 6: green
- Terminal 7: brown/white
- Terminal 8: brown

This type of termination takes a little more time than a coax termination, but after you have done a few, it will be second nature. Buy a bag of RJ45 connectors to keep on hand. Soon you will be making up these cables for neighbors and at work. Terminating Ethernet lines is the most important single skill that is needed to create a data network in the home.

## USB Cables

For connecting peripheral devices to a computer, Universal Serial Bus (USB) cabling is a great advance over older methods, including serial ports, telephone-line cords, and similar expedients. Introduced in the 1990s, USB cabling is now a widely used cross-platform (Mac and PC) medium becaue of its high-speed connectivity, ease of use, and reliability. Modern computers have multiple USB ports. We have all seen the familiar logo, shown in Figure 11-9. If you have more devices than ports, just add a USB hub.

Many computer peripherals, such as keyboards and mice, have permanently attached USB cords that simply plug into any one of the computer's USB ports. There is no need to power down or restart the machine. A grounded conductor makes contact before the data pin, so there is no risk of damage from static charge.

Besides printers, keyboards, and mice, USB cabling is used to connect joy-sticks, flight yokes, webcams, digital cameras, modems, speakers, telephones, TVs, external data storage devices, and network connections. If a new device is plugged

**FIGURE 11-9**   The ubiquitous USB logo, indicating a port that will accept a USB cable connector.

into a computer or hub via USB cable, the computer's operating system detects it, requests the driver disk, and activates it.

The two ends of a USB cord are not the same. The upstream end has an A connector with a corresponding socket at the computer. The downstream end, if it is not built into the device, has a B connector. These ends are completely different, so it is not possible to plug the cord in the wrong way.

As many as 127 devices can connect to a computer through existing USB ports or line-powered USB hubs. Besides data, USB connections can supply power to the device, up to close to 1 amp at 5 volts. For the home crafter-electrician, USB is good news. Just plug it in, and you are ready to go to work.

Before you undertake a full-scale networking project, just to gain experience and confidence, we'll start with some simple exercises. These can be simple thought experiments, if you don't want to expend the time and materials.

## Telephone Extensions

The simplest form of communication network is a telephone line extension. You have probably done one or more of them. The least sophisticated way to do this would be to purchase extra long line cords with modular RJ11 ends and extend the line using those little couplings found in discount stores. But good telephone work involves a little more planning. Always the telephone wire should be concealed, not just stapled along a baseboard. (If the wall is concrete and there is no way to conceal the telephone wire, consider Wiremold.)

In new construction, the telephone company will run the line to an interface box mounted on the outside wall. One section of this enclosure is accessible to the user, and it contains provisions to connect at least two lines using insulation-displacement terminations. It also contains a modular socket for test purposes. The utility grounds its system at this interface box by running a bare or insulated copper wire to the intersystem bonding terminal provided as part of the electrical installation. If this piece of hardware is not in place, the utility will seek out some other available ground connection. The meter socket enclosure is a possibility.

If the electrical service consists of an underground lateral, polyvinyl chloride (PVC) raceway for the telephone service should have been installed in the same trench. If this is the case, a vertical PVC riser, connected to a 90-degree sweep, should have been left as a short above-ground stub with a pull rope inside. Telephone utility policies vary, but the usual procedure is that the utility pulls a two-pair telephone line through the customer-installed raceway and connects it at the interface. It is the owner's task to complete the installation.

To avoid wire clutter on the outside wall, bring the two pairs through the back of the interface and through a hole in the wall into the building. The telephone company can configure these two pairs as either separate lines with different telephone numbers, two lines on the same extension, or one line with a spare. For this and all other inside telephone wiring, it is recommended that Category 5e cable be used as opposed to two-pair telephone line. The advantages are

- You already have Category 5e cable for the data network, and it saves buying separate telephone wire with an additional cutoff.
- Category 5e contains four pairs, so in every telephone line there are spare pairs that can be used later if extra lines are desired. If a nail pierces a conductor behind finished wallboard, you can swing over to an unused pair at both ends.
- Category 5e solid conductors can be stripped and coiled under a terminal screw or punched down using a special tool into an insulation-displacement terminal.
- Used or creased pieces of Category 5e that are judged to be unsuitable for data transmission can be used as telephone wire because the low audio frequencies are less demanding.

The Category 5e from the telephone interface, once it has been brought into the house, can be run to a central location where all the separate telephone lines will originate. This can be on a wall in the basement, in a utility closet, or in a suitable location of your choosing. It should not be close to ac wiring, inductive loads such as a motor, UPS, fluorescent ballast, or other source of interference. They generate harmonics, and at the very least, a faint but persistent hum may be heard on all telephone lines.

The Category 5e that has entered the house should terminate in a dedicated telephone enclosure. This can be as simple as a 4 × 4 junction box or, for more elaborate installations, a punch-down board. A small telephone cabinet with a door that opens will keep out the dust and yet allow easy access.

Category 5e conductors used in telephone lines (but not in data lines) can be joined in a variety of ways. A very primitive method that works but will be found unsatisfactory in the long run is to merely twist the stripped wires together and tape them. Any jostling will cause these joints to work lose, and if there is a great number of them, troubleshooting will become an ordeal. A far better method is to use small blue wire nuts. Strip $\frac{3}{8}$ inch of insulation from the end of each conductor. Lay them side by side, being sure that the ends are flush. Do not twist the wires. Let the wire nut do the twisting. Turn the wire nut clockwise until a definite resistance is felt. Do not tape the wire nut. As long as the ends were flush to begin with and the proper amount of insulation was removed, this type of splice is durable and trouble-free.

The *NEC* does not require an enclosure for this type of low-voltage splice, but it is more professional to put it in an enclosure with a cover that is accessible and labeled. The wire nuts should be positioned with the openings down so that any moisture will drain.

The next step up for a telephone line splice is the silicone-filled bug. This is an insulation-displacement method. The Category 5e (or telephone) wires, without being stripped, are inserted into the bug so that they are felt to bottom out. A special crimper, with two stops, one to ensure correct positioning and the other to regulate the amount of compression, is used to complete the connection. Silicone is observed to ooze out of the holes around the conductors. This is the most professional way to splice telephone wires. It is quick, waterproof, and resistant to vibration. Because the wires are not stripped, there is no need to worry about copper exposed on the outside.

To emphasize, these splicing methods cannot be used for data lines. If a data line is found to be defective, it is customary to remove it and install a new cable that is free of splices. If a splice or tap has to be made in a data line, an Ethernet hub or better must be used. This presupposes an ac power source nearby, unless a battery-powered unit is to be used.

Unlike electrical branch circuits, which are ordinarily daisy-chained, individual phone jacks are best run back to the source in what is known as *star topology*. This is feasible because Category 5e is far less expensive that 12 AWG Romex.

Category 5e cable lends itself to premises telephone wiring. The conductors may be punched down or coiled and tightened under a terminal screw. When bringing Category 5e into a 4 × 4 metal box, remove a knockout and insert a rubber knockout blank that has a small hole drilled in it. Either that, or you can make a small X with a utility knife. The Category 5e can be brought into the enclosure using this fitting, and it will never ground out against the metal or fall out of place.

One of the benefits of Category 5e in telephone work is that there are extra spare pairs. After stripping the jacket, leave them full length, and coil them around the jacket. They will be there for future use, and they won't add to enclosure fill. In addition, they make a strain relief for the cable where it enters the jack or enclosure.

For telephone cabling, Category 5e can be spliced to two-pair telephone wire. If you are using Category 5 in a building or addition that has existing telephone wires, they should be spliced consistently so that future circuit tracing and troubleshooting are facilitated. The color codes are different and should be matched up as shown in Table 11-1.

TABLE 11-1    Color Codes for Telephone Lines

| Telephone Line | Category 5 | Old Terminology | Telephone Wire |
|---|---|---|---|
| Pair 1 | Blue | Ring | Red |
| | Blue/white | Tip | Green |
| Pair 2 | Orange | Ring | Yellow |
| | Orange/white | Tip | Black |
| Pair 3 | Green | Ring | — |
| | Green/white | Tip | — |
| Pair 4 | Brown | Ring | — |
| | Brown/white | Tip | — |

# Tools for Telephone Work

In addition to the ordinary electrician, carpentry, and mechanical tools discussed in Chapter 6, some specialized tools are needed for telephone installation, trouble-shooting, and repair. All of them are not essential for a home job, and they can be acquired on an as-needed basis.

For a simple single-extension installation, you won't need much in the way of test tools. Chances are that you will complete the hookup, and it will work fine. Only when part of a complex system is not working in a large building, especially if it is old and the cabling layout is uncertain, would specialized testing equipment be needed. If you are likely to be doing more telephone work in the future, now is the time to get equipped, starting with the less expensive and more needed test tools.

You already have the most basic test tool used in telephone work. It is a simple touch-tone phone with a modular line cord. As an accessory, wire two flexible leads with small alligator clips or wire ends into a telephone jack that has a modular socket, as shown in Figure 11-10. Then you can use your telephone as a test set to access telephone service and test for dial tone where there is no modular connection.

An upgrade would be a professional test set. Less expensive models sell for under $100, whereas full-featured instruments that combine data-networking capability are far costlier. The test set (*butt set*) is, for telephone work, the real bread and butter tool used by telephone technicians over and over in the course of a day. There is an inexpensive test light with a modular plug that indicates ringing and dial tone, but you probably won't use it if you have either the homemade or professional test set described earlier.

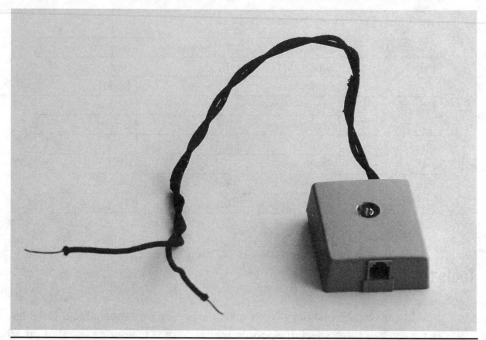

**Figure 11-10**    Simple homemade test accessory to go with a touch-tone telephone.

The tone generator (*toner*) is a convenient little tool that greatly expands the capability of the test set. Connected to a telephone line, it generates a distinctive two-tone audio signal that can be heard over a test set or telephone. With it connected in place, you can test the integrity of the phone line, jack, line cord, and phone. The toner is equipped with a modular socket, short line cord, indicator light, and switch that must be turned off between uses so that the 9-volt battery will not be drained.

The *wand* is used along with the toner. It is a noncontact tool that picks up the audio tone when the pointed probe is brought near either member of the live pair. If there is a large mass of bundled pairs, it permits the user to quickly identify the energized conductor without need of hooking onto each pair individually.

Lacking these specialized tools, it is possible to diagnose faulty extensions by means of an electrician's multimeter, but this method is less convenient. For example, you can take ohm readings on a pair to see if it is shorted, but to test for an open line, you would need to go to the other end of the line and short out the pair with a jumper. After taking the reading, it is necessary to go back and remove the shunt.

## Telephone Installation and Repair Tools

As for telephone installation tools, besides the ordinary carpentry, mechanical, and electrician's tool, these items are needed:

- **A long, thin wood bit, $\frac{3}{8} \times 30$ inches.** Additional bit extensions, fastened with Allen-head setscrews, are helpful in some situations.
- **A telephone wire stapler.** This resembles a standard stapler used for hanging foil- or paper-faced fiberglass insulation, but it has a specialized head suitable for rounded telephone wire staples.
- **A punch-down tool.** This is used to make terminations in insulation-displacement connectors, punch-down blocks, patch panels, and the like.

Professional telephone service technicians have much more equipment, and some of it costs thousands of dollars. But the items just mentioned almost always will suffice, and where they do not, your ingenuity will create a work-around.

## Diagnosing Some Common Telephone Problems

Because a telephone contains semiconductors, a constant voltage is required to provide bias. This is true even when the phone is hung up in the off state. The voltage for a phone in this dormant state is 6 volts dc. When the receiver is taken off the hook, the level rises to 50 volts dc. Ring voltage goes up to 130 volts at 50 Hz.

Power-supply voltage along with an audio or audiovideo signal is quite common in many types of communications systems. If the electronic equipment is not battery powered, has no power cord, and contains semiconductors, it is usually network powered. If this power is ac, the local equipment will have a power supply, usually including a transformer. This is true for cable TV (CATV), where the standard is 60 volts ac. This makes it possible to trace CATV lines, usually coaxial cable, both indoors and outside, using standard test equipment, as shown in Figure 11-11. If the voltage is dc, the premises equipment will not contain a transformer. In a telephone system, there is no conflict between the power-supply voltage and the signal because they are easily separated by means of an electronic filter consisting of capacitive and inductive elements.

The power-supply voltage can be used for troubleshooting using a multimeter. Where it is not present, there will be no audio signal, and if it were there, it

**FIGURE 11-11**   CATV lines can be checked using a multimeter to detect the supply voltage. Also, a small TV or field-strength meter can be used to determine the presence and quality of the signal.

wouldn't do any good. But it is better to use a test set to get a good idea of what is going on because this instrument will provide information about the quality of the sound and let you know if there is noise on the line.

The most frequent complaint about a phone that is not working is that there is no dial tone. This is the continuous ac hum that is generated by the phone company, indicating that there is an active connection back to the central office. Very often if there is no dial tone and no sound is heard at the receiver, it is the telephone that is at fault. There are three quick and easy ways to determine if this is the case:

- Connect a known good phone to the jack.
- Connect the phone in question to a known good jack.
- Remove the cover of the jack, and connect your test set to the terminals.

If your finding is that the phone is at fault, don't be quick to discard it. Over 50 percent of the time, it is the line cord that has become defective. A sure sign is if the dial tone comes and goes when you wiggle the cord close to either end. Disconnect the two ends from the jack and phone, and try a new line cord. Also, sometimes a one-line cord has been plugged into the wrong socket at the jack. Perhaps at one time a separate line, now discontinued, was provided for a dial-up Internet connection, and for some reason, the line cord was unplugged and then reconnected to the wrong modular socket. If this is not the case, putting in a new line cord often clears up the problem, but be aware that a brand-new line cord could be defective. Try it on a known good phone connection.

## Bad Receiver Cord

The receiver, the part of the phone that contains the earphone and microphone, is connected to the phone body by means of a receiver cord, which is a tightly coiled line that matches the color of the phone. It also has modular connectors at both ends, but they are smaller and not compatible with the line-cord modular sockets. When there is no dial tone, this cord is sometimes at fault, and it can be easily checked by substitution in the same manner as the line cord. Also, the receiver may have gone bad, and it also can be checked by substitution. A receiver from a different model phone will work electrically, but it may not sit in the cradle and hang up properly, and for most applications, you also will want the color to match.

The phone may have ac power provided by a thin power cord that plugs into a nearby wall receptacle. There will be a small in-line rectangular transformer right at the wall receptacle. If the phone has ac power, there will be a power light. The purpose of this ac power is to enable certain accessories such as an alphanumeric display and other bells and whistles. If the ac power fails due to a utility outage or tripped breaker in the house, the phone will still function without the accessories, although the power light and accessories will be out. Very often this condition is caused by the plug becoming loose because of the weight of the transformer.

If there is power at the receptacle but not at the phone, it is likely that the transformer has failed. A replacement must be the same voltage as shown on the transformer nameplate. A mistake is unlikely because different voltages have different plug configurations, which are not mutually compatible. But beware that the transformer output impedance should be an exact match. If the cord or transformer is overloaded, there could be a fire hazard. The transformer will be felt to run warm, even when the phone is not in use, and if windings inside should become shorted, it could become hot, so the area should be kept clear of combustible materials.

## Can't Dial Out?

Occasionally, a phone will have dial tone and good sound, but it will be unable to dial out. This could be the fault of the utility, and that can be determined by phone substitution. If the defect is in the phone, the keypad has likely failed. If it is an expensive desk phone with an alphanumeric readout, you may want to open it up and consider replacing the keypad. This is feasible, and for many models, repair parts are available.

If the phone, including receiver and cords, turns out to be good, unplug the line cord from the jack and listen to the signal at that point, using either a good phone plugged into the modular socket or a test set at the jack terminals. Most jacks have two pairs of terminals to accommodate a two-line phone. You can determine which pair is relevant by looking at the wiring, incoming and outgoing.

The original complaint by the user could be no dial tone, intermittent dial tone and sound with intermittent ability to call out, noise on the line, or ac hum. Frequently, the fault is in the jack. One of the fine wires can be broken, or it may have come loose at the terminal. This is a particular problem when there is more than one wire under a screw. It is usually possible to wiggle wires in the jack while listening at the receiver and locate any problem that is in that area.

A professional test set has alligator clips fitted with sharp needle-like points that will penetrate the wire insulation and contact the copper conductor. Without stripping insulation, you can check for a dial tone where the telephone line enters the jack.

A telephone will appear dead if the line is either shorted or open. If there is a solid short or both members of the pair are open, the phone will be completely silent. If one member of the pair is open and the other is intact, there may be a very faint ac hum or faint background noise. If any phone in the building is shorted out, it will pull down all the other phones, so in such a situation, it is sometimes best to unplug every phone (including credit-card readers and the like) to see if that restores service and then start plugging them in one at a time.

If there is a long stretch of telephone cable that is concealed by wall finish and tests at both ends indicate that it is at fault, the best way to proceed could be to abandon the line and run new cable. It may be acceptable to relocate the jack.

## Noise on the Line

If there is noise on the line, it may be caused by a partial fault between members of a pair or a partial fault to ground of a single wire. This can be subtle and difficult to find by visual inspection. If a telephone line is buried, even in PVC conduit going to an outbuilding, that can be the source of the noise. If you disconnect both members of the pair at the near end and the noise goes away, you have located the fault. Buried telephone lines are particularly vulnerable to this sort of fault where they come out of the ground at either end.

Outdoor aerial cable is also a frequent source of noise. If the noise appears during rain or windy conditions, that is an important clue. Years ago, telephone lines were customarily run on the outside of a house stapled to the siding to penetrate the wall at individual jack locations. This old cabling can come loose over the years, and the wind will cause it to slap against the wall, damaging the insulation and eventually making for noise. The bad spot in a noisy telephone line can be difficult to locate because a very small nick in the insulation can cause the problem. An ac hum can be caused by a telephone line that is located close to a power line or nonlinear (inductive) load. Opening circuit breakers individually can be helpful in locating the source.

# CHAPTER 12

# Home Automation and Beyond

For a long time, home automation was a disorganized set of fantasies for science fiction writers, highly motivated underachieving hobbyists, and owners of exotic mansions for whom cost was not as issue. Now, suddenly, all sorts of off-the-shelf software and hardware, as shown in Figure 12-1, are readily available at moderate cost for any home crafter-electrician who is willing and able to do the installation.

In Chapter 11, we saw how to install coaxial, USB, and Ethernet cabling, and by now, you should be somewhere between competent and adept at running branch circuits and powering up loads, including stepper and servomotors.

## Before You Start

As a reminder, if you are currently wiring a new home or addition, be sure to include a neutral with every switch loop. The *National Electrical Code* (*NEC*) requires it, and it is necessary when you decide to pull out the old-world switch and put in a smart device. Also, if you are in the process of roughing in wiring for a home or addition that may at some time in the future become automated, consider using plastic as opposed to metal wall boxes. This is important for certain types of home automation where the devices communicate wirelessly because the grounded metal enclosures may inhibit radiofrequency (RF) transmission among devices.

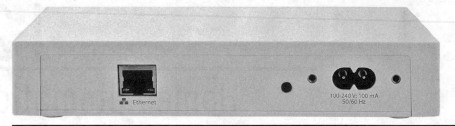

FIGURE 12-1    A home automation hub, the heart of the system. (*Courtesy of Insteon.*)

If you think about it, homes have been automated to a certain extent for a long time. *Automatic* means self-acting without human intervention. A simple thermostat that controls oil, gas, or electric heat is an example, and so is the thermostat that controls a hot-water heater. The quick-recovery (or tankless) hot-water heater is an interesting subspecies that is also entirely automatic.

In its simplest form, an automatic device consists of two parts—a sensor, shown in Figure 12-2, in a low-current loop and an actuator connected to a higher-current electrical supply. The power is switched on or off by means of a mechanical relay or solid-state switch. The arrangement is good because it is not

FIGURE 12-2    A leak detector can prevent extensive water damage. (*Courtesy of Insteon.*)

necessary to run the high-current power over to and through the sensor. In our earlier example, there is such a thing as an in-line thermostat where the entire power that runs the actuator also passes through the sensor. This is a less sophisticated arrangement and generally is avoided.

Consider the common doorbell. It is not an automatic device because a human is needed to push the button. But substitute a motion detector, as shown in Figure 12-3, and you have an automatic device, a simple step toward home automation.

Building automation has been around for many decades in commercial and industrial facilities. The fire alarm system in a large hotel is an amazingly intricate and robust engineering marvel. It saves lives by combining nineteenth-century sprinkler technology with the latest computer control capability. Ever vigilant, it constantly monitors its own circuitry, including two redundant automatic telephone ties to the fire department and phase 1 and phase 2 control of the elevators so that they will automatically bring passengers safely to a floor that is not burning and then become available for the use of firefighters.

On the factory floor, since the 1960s, programmable-logic controllers (PLCs) have automated production to such an extent that robots can assemble, weld,

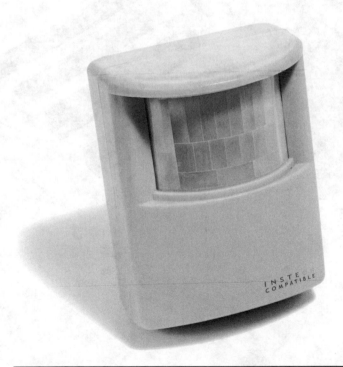

**Figure 12-3**  A motion detector can be used to switch on lights as you progress from room to room. (*Courtesy of Insteon.*)

paint, and, in short, do whatever it takes to build cars, trucks, and other products. When next year's models come out, the entire manufacturing process can be reconfigured by means of laptop computers using onscreen ladder diagrams.

The point is that this technology is nothing new. The only problem is that even the simplest of such systems, as described earlier, used to cost millions of dollars. Today, suddenly, it has all become affordable.

Almost any object or process in a home is capable of being automated. An example is shown in Figure 12-4. Automation just means removing the human element from the loop. Some of the things that can participate in home automation are

- Appliances
- Windows
- Doors, including garage doors
- Loud speakers

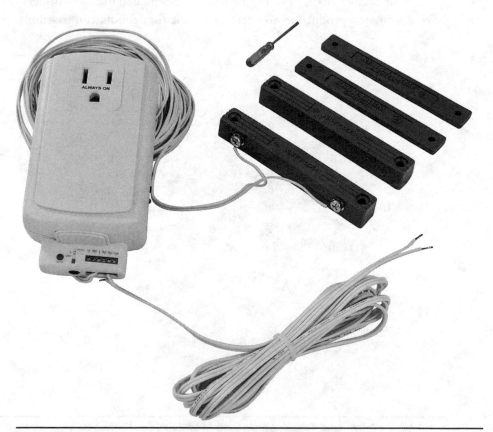

FIGURE 12-4    A garage door controller with monitoring kit. (*Courtesy of Insteon.*)

- Clocks
- Pet feeders
- Window shades
- Irrigation
- Computers
- Lamps and room lighting
- Surveillance cameras
- Telephones
- Radios, TVs, and any audio or video equipment
- Premises electrical system
- Heating, air conditioning
- Fire alarm

Some home automation units can be active, having the ability to turn on the heat or drain a swimming pool, for example, whereas others may be passive, such as a surveillance camera, as shown in Figure 12-5, that looks out across a yard. The output can be made available via a password-protected website so that it can be viewed anywhere that there is a computer with an Internet connection.

There are numerous competing vendors offering home automation products, and it should not be assumed that they are compatible. Differing communication

AUDIO  OUT  IN  DC 5V 2A

**FIGURE 12-5**    A wireless surveillance camera monitors the area. (*Courtesy of Insteon.*)

protocols, voltage levels, cabling requirements, and mechanical setups make for a Tower of Babel situation in the world of home automation.

For this and other reasons, it is best to decide at the outset on a single product line and stick to it insofar as possible. For most of us, the choice will be driven by cost, but there are other factors worth considering as well. For one thing, you want an outfit that offers a broad array of devices, such as the wireless-controlled receptacle shown in Figure 12-6, with the ability to start with just the hub and add items as desired, as opposed to being given a choice of predetermined packages. You want a choice of wireless or hardwired, if required by the system you are considering, with the flexibility to choose between the two in case your ideas change. And you want the ability to mix systems without paying a penalty in added cost.

If you are building and wiring a new house and considering including the home automation dimension, before hanging the wall and ceiling finish materials, you need to consider the cable versus wireless options. In the past, home automation systems were complex, unwieldy, and, as a result, far too expensive for widespread use. Since 2000, a new home automation suite has been developed by Insteon. Starting at the company's website (www.insteon.com), the home crafter-

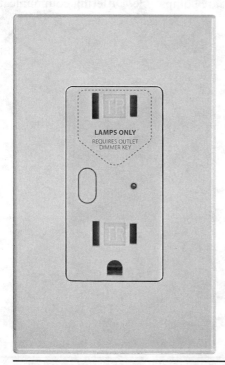

**Figure 12-6**   A wireless-controlled receptacle. (*Courtesy of Insteon.*)

electrician can learn how to put together a low-cost system that is reliable and offers all the home automation features that one would desire.

It is possible to begin with the hub, which controls and integrates all system components, and then in an incremental fashion add what is desired for a given installation. No hardwiring is required. Communication between the hub and all devices takes place wirelessly via radio communication at 15 MHz and simultaneously, as a redundant means of communication, over the existing branch-circuit wiring. Crystals in each Insteon device, such as the one shown in Figure 12-7, generate a power-line carrier frequency of 131.65 KHz, which is modulated as needed to convey information in packets between the hub and members of the network—all home automation devices. These signals coexist with the 60-Hz electrical power carried by the electrical lines to the detriment of neither.

The range is about 150 feet. Each device, as it is added, becomes a receiver and retransmitter, so there should never be a problem in extending coverage throughout the building. From the point of view of the home crafter-electrician, the whole thing is plug and play. No advanced wiring or programming skills are needed.

Insteon devices, through the hub, can communicate with the Internet, smart phones, tablets, and desktop and laptop computers. Connectivity is achieved through a dedicated serial interface such as a USB, RS232, or Ethernet connection, as shown in Figure 12-8. Just run the cable to your router or modem, and you're all set.

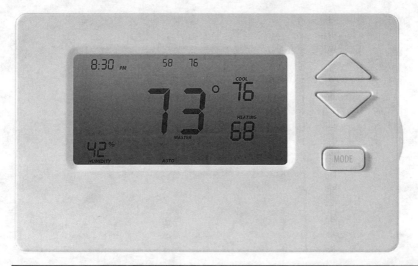

FIGURE 12-7    A thermostat with humidity display. (*Courtesy of Insteon.*)

**FIGURE 12-8**    A home automation receptacle with an Ethernet connection. (*Courtesy of Insteon.*)

CHAPTER **13**

# Backup
# Power
# Systems

T he utility electrical supply is vulnerable to interruption, momentary or long term, possibly lasting for days. Adverse weather, flooding, system overload, terrorist attacks, power-pole breakage due to intoxicated or texting motorists, and all sorts of other catastrophes can disrupt the utility connection so that the home is without power.

## Backup Power for the Home

You can meet this challenge by installing a backup power system. It is required for health-care facilities, where continuity of power is essential for life support and ongoing operations. Schools, hotels, office buildings, and industrial facilities generally have backup power, but for the home, it is usually optional. A small propane-fueled generator in a dedicated building is shown in Figure 13-1.

If you are located in an area that experiences cold weather in the winter and you depend on electricity to heat your house, you will definitely want backup power. If there is an outage that lasts more than a few hours and it coincides with a dip in the temperature, the cost of repairing burst water pipes in finished walls and ceilings may well exceed the cost of installed backup power. Another consideration is the ongoing need for electricity to power the water pump. (If you have a municipal or gravity system, this is not a problem.)

After evaluating your needs, if you decide to put in a backup system, there are a few decisions to be made. A very low-end backup power system would consist

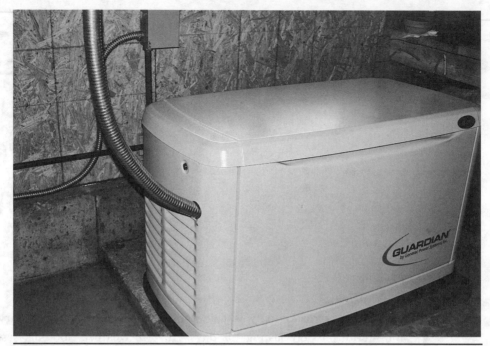

**FIGURE 13-1**    A small high-end generator.

of nothing more than a gasoline generator that is rolled out of the garage and started up outside to supply limited plug-in capability during an outage.

At the other end of the scale is a large diesel-powered generator in a permanent, dedicated building, sized to carry the full load, with automatic starting and an automatic transfer switch. There are intermediate solutions that are less costly, require a human operator, or power less than the full load. In this chapter, we will look at some of the options. Throughout, it is essential to consider safety above all. Designing and building a backup power system involve some serious electrical work, so it is necessary to research the whole area thoroughly. Don't hesitate to bring in a licensed electrician who specializes in such installations if you are unsure of some of the details.

## A Transfer Switch Is Needed

First and foremost, we must emphasize that the electrical energy from the backup power supply must not be allowed to backfeed into the utility distribution system. If the backup power is connected to the entrance panel in order to supply premises loads or indeed if it is connected to any part of the electrical system inside or outside the building, there is the distinct possibility that it will backfeed through the service entrance conductors into the utility lines. Moreover, because this power is applied to the output lugs of the utility pole- or pad-mounted transformer, this normally step-down device will become a step-up transformer, energizing the power line to a high voltage level and endangering the lives of utility workers who have every reason to believe that they are working on lines that are at ground potential.

It is not sufficient to rely on the main breaker to provide the needed isolation. The problem is that while the backup power is running, someone may come along and reset that switch. The isolation has to be absolutely failsafe because human lives depend on it.

Totally reliable isolation of the backup power from the utility transformer and lines is provided by a *transfer switch*. A schematic diagram of such a switch is shown in Figure 13-2.

For a 120/240-volt single-phase service, this is essentially a double-pole, double-throw heavy-duty switch that is sized to carry the electrical load. It is a physical impossibility, when the transfer switch has been wired in place properly, for the entrance panel to be connected simultaneously to the backup power source and to the service-entrance conductors.

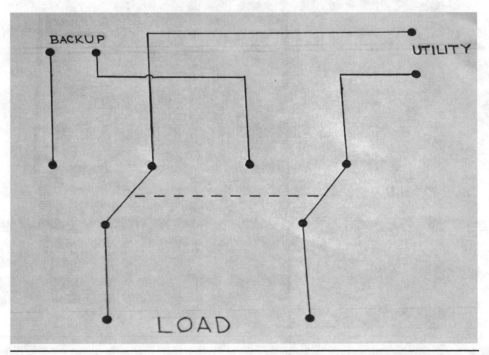

**FIGURE 13-2**   This transfer switch schematic shows the load (premises wiring) connected to the utility supply.

The transfer switch is usually in a separate enclosure from the entrance panel. The enclosure may be for indoor or outdoor use, and it may throw the entire premises load over to the backup power source or it may connect only selected circuits. The second of these types, shown in Figure 13-3, is often used because it is less expensive, and quite often, the backup power unit is not capable of powering the entire premises load. To attempt to do so would either stall the engine or trip an overcurrent device.

An automatic transfer switch may include circuitry that detects a utility outage. Ignoring short-duration transients, it powers up the backup unit, which includes a starting battery and charging system. The backup unit is allowed to run a prescribed amount of time so that the engine warms up and output voltage stabilizes, whereupon the transfer switch is automatically thrown, and the building is powered up.

**FIGURE 13-3** The automatic transfer switch often performs other functions, such as sensing a utility outage, starting the generator, and returning to utility power after the outage has ended.

After utility power is restored and has stabilized, the transfer switch reconnects the entrance panel to the normal utility supply. The backup power-source engine is allowed to run a minute or so to cool down, and then it shuts down.

If only selected circuits are to be connected to the backup power source, they should be chosen so that essential loads have power. The electrical power necessary to heat the building, selected lighting, a motorized overhead garage door, water pump, any medical or life-support equipment, and other loads of your choosing should be included.

> ### Backup Power and the *National Electrical Code* (*NEC*)
>
> The 2014 *NEC* covers emergency systems, legally required standby systems, optional standby systems, and critical-operations power systems. Residential backup power systems are covered in Article 702, "Optional Standby Systems." If you are planning to build one of these, go through Article 702 to ensure that your installation is fully compliant. Essential aspects of the optional standby system are the presence of a transfer switch, effective grounding, and overcurrent protection.

## Fuel Options

The backup power supply should be carefully sized with an adequate safety margin to supply the selected loads. The type of power plant should be appropriate to the task. Gasoline is to be avoided for all but the smallest installations because handling gasoline can be hazardous in terms of fire.

Diesel fuel is less volatile. For a large building that is to be fully powered, it may make sense to have a diesel power plant with a free-standing fuel tank. A disadvantage of diesel is that diesel engines are difficult to start in cold weather. An engine heater will draw as much as 3,000 watts, so it is costly to keep it heated full time in winter waiting for an outage.

Many installations are run on bottled liquid propane gas (LPG), and this is found to be quite satisfactory. Moreover, a gasoline engine can be converted to run on LPG.

Every effort must be made to protect the occupants from exhaust fumes. The power plant, with a suitable housing, can be located outdoors as long as the exhaust is guarded so that there is no chance children or others will breathe the fumes or get burned by contact with the exhaust system. The power plant can be in a well-ventilated dedicated building. It must be exhausted to the outside, including the crankcase vent, and clearances from wood and other combustible materials must be maintained.

A common residential installation is an LPG-fired engine powering a 120/240-volt single-phase alternator. It is permanently mounted on a slab located a few feet from the building, not near a window or door. It is solidly bolted to the concrete slab. The electric line to the building is in polyvinyl chloride (PVC) conduit, which goes underground to the house. The LPG supply line is also buried. In forming up for the slab, it is necessary to provide electrical conduit for the power and any control conductors and a sleeve for the fuel supply.

# CHAPTER 14

# Alternate Power Systems: Wind, Solar, and Fuel Cells

E very year, our planet is getting a little warmer. For those who live in cold areas, this may seem beneficial, but it is actually an enormous problem for everyone, and it goes way beyond discomfort. When ocean waters are heated, vast amounts of energy are injected into offshore weather systems. Together with glacial melting, which is going to raise sea levels, these events will conspire to take away our coastal lands.

A very few degrees of warming will disrupt established patterns of agriculture, fishing, and all sorts of human activities. Economic dislocations and widespread poverty will follow. We have to stop burning carbon to heat our buildings and power our vehicles—the sooner the better. Changes must be made. For now, it is looking like the changes will be incremental and highly localized.

## Some Answers

There has been a lot of interest lately in alternate (i.e., non-carbon-based) energy systems, and the home crafter-electrician is very well positioned to participate. Tax and utility incentives in many locations make it affordable to build onto our homes the means to heat them, provide electricity, and power our vehicles. The technology exists to do this right now.

The fantasy of perpetual motion is not an answer. If in the future some quirk in quantum mechanics or a way to reverse the arrow of time and make entropy work for us should appear, that would change the situation dramatically, but for the time being, we have to work with what we have, which is two very abundant energy sources that are everywhere on Earth—wind and sun.

## Wind

Many of us have entertained the notion of fitting some sort of rotor to a surplus automotive alternator, building a wind turbine, and putting a stop to electric bills forever. This is part fantasy and part reality. The power that is produced within, let's say, a Delco-Remy alternator is known as *wild alternating current* (ac). Because of the circular rotation of the armature, the waveform is an elegant sine wave that is free of harmonics and noise. But voltage and frequency fluctuate rapidly as the speed of rotation changes. The three-phase ac is rectified by means of a network consisting of six diodes inside the alternator so that the output is regulated direct current (dc) that is perfect for charging an automotive battery.

This sort of alternator can be used as the basis for a wind generator, but there are a few problems. First of all, the revolutions per minute produced by a wind-driven rotor are way too low to operate the alternator. In order to drive an automotive alternator at a speed where it would produce its rated power, a set of pulleys or a geared transmission would be necessary to increase the revolutions per minute.

It is possible to rewind the stator, replacing the existing windings with more turns of a smaller gauge wire. This is a job for a specialized motor shop unless you plan to dedicate yourself to a big task with little return. The total output still would be far too low for even a single household. Additionally, the automotive alternator would not have a weatherproof housing. Overall, the potential for building a usable wind generator from an automotive-type alternator is low.

As electric cars including hybrids become more prevalent, older cars and those that have been wrecked will end their days in junkyards. Many electric cars employ regenerative braking, whereby the car motor is used to slow the vehicle and simultaneously charge the battery bank as opposed to dissipating the energy as waste heat from the friction of brake pads and wheel hubs. As designs become more diverse and old electric vehicles are put out of use, it is likely that some of these motor-generators will find second lives as wind generators.

Meanwhile, the option of using factory-built wind-turbine equipment, as shown in Figure 14-1, is very appealing. Many of these highly engineered power plants are well suited for residential use. With public and private utility incentives, these installations are quite feasible.

Before deciding on a model, it is essential to evaluate the feasibility of the project and, if it is a go, to size the turbine. To do this, you will need to choose the best location. Naturally, it should be near enough to the house so that a long

**FIGURE 14-1**    A wind turbine configured to produce electricity.

transmission line will not be needed. Also, it is very desirable to have the wind generator visible from a house window. The tower should be far enough from the house and any other buildings that the airflow is not disrupted because this makes for turbulence and less useful wind power. Similarly, nearby trees or other objects will reduce output. (Don't forget that trees will continue to grow during the life of the installation.) A tall tower will help to overcome these limitations, and if there is a knoll or upslope location for the wind turbine, this is a plus.

## Choosing the Site

A temporary pole with an anemometer will help in evaluating the site. A wind-direction indicator is also useful. Sustained unvarying winds that do not shift direction constantly are far better than brief, powerful gusts.

The wind-turbine manufacturer will have documentation that includes wind-speed requirements for different power outputs, so you will want to evaluate your needs and choose the appropriate model. The height of the tower is an important variable. An excessively high tower will be an expensive and inconvenient burden, but on the other hand, the wind quality in terms of sustained speed and lack of turbulence improves with distance above grade.

You should contact the local planning board or regulatory authority early in the planning stage to determine whether there is a height limitation, property-line setback, or minimum lot size that needs to be observed. Tax and electric-rate incentives should be investigated at this stage, and if it is a cogeneration project, the utility should be contacted in advance.

In most jurisdictions, the *National Electrical Code* (NEC) or other code governs every detail of a nonutility wind-generator installation. Previous *NEC* editions covered only small wind electric generators, defined as up to and including 100 kW of output. This limit was removed from the 2014 *NEC*, so now any nonutility installation must comply. In the early planning stages, you should carefully review *NEC* Article 694, "Wind Electrical Systems," to make sure that nothing is neglected. As an overview of this article, here are some definitions:

- *Charge controller* is equipment that controls dc voltage or current or both used to charge batteries. Where a battery backup system is used, a good charge controller will contribute to a safe and reliable installation. It prevents overcharging, which can damage batteries and release hydrogen with a risk of explosion and fire.

- *Diversion charge controller*, in conjunction with the charge controller, further protects the battery bank from the risk of overcharging. When the batteries are fully charged and output exceeds demand, excess power is diverted to a dummy load.

- *Guy* is a cable that braces a wind turbine tower so that it will not buckle during heavy winds. Some towers are sufficiently strong and have adequate concrete foundations that guys are not needed.

- *Inverter output circuit* consists of the conductors between the inverter and service equipment or another power source.

- *Maximum output power* is the greatest 1-minute average power output a wind turbine produces in normal steady-state operation. Instantaneous power output will be higher at times. This amount determines the size of the inverter, rectifier, and associated wiring.

- *Maximum voltage* is the highest voltage the wind turbine can produce, including open-circuit conditions. This figure also plays a role in selection of the inverter and rectifier.

- *Nacelle* is the enclosure that houses the wind turbine, including the alternator but not the rotor.
- *Rated power* is the output power of the wind turbine at a wind speed of 24.6 mi/h.
- *Wind-turbine output circuit* consists of the circuit conductors between the internal components of the wind turbine and other equipment.

*NEC* Section 694.7, "Installation," states that wind-turbine systems are to be installed only by qualified persons. Because there is no homeowner exemption for this requirement, as there is for electrical licensing, the home crafter-electrician will have to become qualified. This is an important mandate, and it is in your interest to become qualified before you perform a wind-power installation. An exact mechanism is not specified. In *NEC* Article 100, "Definitions," *qualified person* is defined as anyone who has skills and knowledge related to the construction and operation of the electrical equipment and installation and has received safety training to recognize and avoid the hazards involved.

An installation requirement for wind-turbine systems is that a surge-protection device is to be placed between the installation and any loads served by the premises electrical system. Also, it is provided that wind-turbine electrical system currents are to be considered continuous. The reasoning behind this requirement is that very often the wind can be expected to blow continuously for over 3 hours.

The inverter output of a stand-alone (i.e., non-utility-connected) wind-turbine electrical system is permitted to supply 120 volts to single-phase, three-wire 120/240-volt service equipment or distribution panels where there are no 240-volt outlets. Multiwire branch circuits are precluded for obvious reasons. For a 120-volt wind generator, this is a common arrangement to permit full use of the box. The one hot, ungrounded conductor from the inverter is connected to both bus bars in the distribution panel.

*NEC* Article 694, Part III, "Disconnecting Means," is crucial to getting it right in wiring a wind-turbine electrical system. This is so because the turbine leads always must be considered live, even when the turbine is not turning and when a blocking diode, shown in Figure 14-2, is in place.

Above all, means must be provided to disconnect all current-carrying conductors of a wind-turbine electrical power source from all other conductors in the building. The disconnecting means is not required to be suitable as service equipment. It is to consist of manually operable switches or circuit breakers that comply with all the following requirements:

- They must be readily accessible.
- They must be manually operable without exposing the user to live parts.

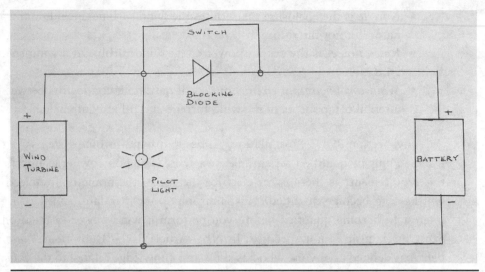

**FIGURE 14-2**    Schematic of a dc wind turbine, blocking diode with motoring switch (for testing purposes), and battery bank.

- They must plainly indicate whether they are open or closed.
- They must have an interrupting rating sufficient for the circuit voltage and fault current available at the terminals.
- Where all terminals are capable of being energized in the open position, a warning sign is required.
- A disconnecting means is to be installed at a readily accessible location or adjacent to the wind-turbine tower on the outside of the building served or inside at the point of entrance of the wind-system conductors.
- A means must be provided to disconnect individual units such as inverter, batteries, and charge controllers from all ungrounded conductors of all sources.

*NEC* Article 694, Part IV, "Wiring Methods," states that all wind-turbine output circuits that are in readily accessible locations and that operate at over 30 volts are to be installed in raceways. Dc wind-turbine output circuits installed inside a building are to be in metal raceways or metal enclosures from their entry into the building to the first readily accessible disconnecting means.

*NEC* Article 694, Part V, "Grounding," provides that exposed non-current-carrying metal parts of towers, turbine nacelles, other equipment, and conductor enclosures are to be connected to an equipment-grounding conductor. A wind tower is to be connected to one or more auxiliary grounding electrodes. This is in addition to the premises electrical grounding system. This auxiliary grounding electrode can be a ground rod, a tie to the foundation rebar, or another acceptable grounding electrode.

These are the principal *NEC* requirements for a nonutility wind-turbine electrical system. Next, we shall turn to the solar photovoltaic (PV) electrical system. After that, we'll consider battery backup and utility cogeneration because these topics are applicable to both wind and solar systems.

## Sun

In some ways, solar power generation, as shown in Figure 14-3, is preferable to wind power generation. For one thing, it lies flat rather than sticking up in the air, so it is less intrusive. With no moving parts, there is less maintenance, and if repair work is needed, there is no need to climb a tower. Often it is easier to find good sun than to find good wind.

Solar-cell prices have been in freefall in recent years, and there is every reason to anticipate that this trend will continue. Ongoing research and economies of scale are coming together to improve efficiency and reduce manufacturing costs, so increasingly it is looking like solar PV power will replace fossil fuels as the dominant energy-producing technology. The principal kind of solar PV cell is crystalline silicon, as opposed to amorphous or thin-film silicon.

FIGURE 14-3   A roof-mounted solar array is unobtrusive.

A solar PV cell is simply a diode that is packaged and manufactured in such a way that when light in a certain range of frequencies strikes the junction, the cell becomes forward biased and produces a steady dc at a low voltage. The power output of a single cell exposed to direct sunlight is low, but a large number of cells can be connected in series and parallel configurations that are capable of producing power limited only by the number of cells. Because a solar cell, electrically, is a simple diode, it has only two leads—an anode and a cathode. Wiring them to make panels and connecting the panels to make an array are accomplished using conventional copper wire splicing and joining techniques, and the output is a two-wire dc of whatever voltage desired, unless it is a center-tap situation, in which case there is a positive pole, negative pole, and neutral that may be grounded.

> ## Thin-Film Solar PV Cells
>
> Thin-film (amorphous) silicon cells are made by depositing lays of photovoltaic material onto a substrate. The end product may be thinner than a sheet of paper. This approach is less efficient than crystalline silicon solar PV cells, but it is less expensive to manufacture, so based on watts per dollar, it looks promising. Research is ongoing to improve the efficiency. Many applications are suggested, such as thin shingles that will serve as a roof covering and a PV electrical source as well.

There is such a thing as a tracking array, powered by one or more servomotors, that moves slowly throughout the day in order to face directly toward the sun at all times. Most arrays, in the interest of simplicity, are stationary, the angle carefully chosen with reference to the site's latitude to maximize output.

A strange thing about solar cells is the way their output is influenced by the ambient temperature. Most physical and chemical reactions accelerate with a rise in temperature, but the output of solar cells decreases as the temperature goes up. For this reason, northern locations, even where there is less sunshine, may be favorable sites for solar PV power generation.

Solar PV installation and repair are generally less hazardous than working on a wind turbine. There is not the high tower to climb, and you don't have to worry about the rotor that could start turning without warning. But, when the sun is shining, the terminals of a PV array are live, beginning when packing material is removed. Today's solar cells produce output even in diffuse ambient light, and it is not practical to locate a disconnect in or on each panel, so there is a potential for shock. Some technicians place an opaque shield over each panel until the assembly is installed and wired to a disconnect, but this solution can be unwieldy and makeshift.

*NEC* Article 690, "Solar PV Systems," resembles Article 694, "Wind Electrical Systems," but there are a few differences due to variations in the electrical characteristics of the two systems. Article 690 begins with definitions. The principal terms, excluding those covered in Article 694, are

- *Alternating-current module* is a complete, environmentally protected unit that consists of solar cells, optics, inverter, and other components, not including the tracker, that is designed to output ac power when exposed to sunlight.
- *Array* is a mechanically integrated assembly of panels or modules, including the support structure and foundation, tracker, and other components as required to form a dc-producing unit.
- *Bipolar photovoltaic array* is a PV array that has two outputs, each having opposite polarity to a common reference point or center tap.
- *Blocking diode* is a diode used to block reverse flow of current into a PV source circuit.
- *Building-integrated photovoltaics* are defined as PV cells, devices, modules, or modular materials that are integrated into the outer surface or structure of a building and serve as the outer protective surface of that building.
- *Module* is a complete, environmentally protected unit consisting of solar cells, optics, and other components, exclusive of the tracker, designed to generate dc power when exposed to sunlight.
- *Monopole subarray* is a PV subarray that has two conductors in the output circuit, one positive and one negative. Two monopole PV arrays are used to form a bipolar PV array.
- *Panel* is a collection of modules mechanically fastened together, wired, and designed to provide a field-installable unit.
- *Photovoltaic output circuit* is defined as circuit conductors between the PV source circuits and the inverter or dc utilization equipment.
- *Photovoltaic power source* is an array or aggregate of arrays that generates dc power at system voltage and current.
- *Photovoltaic source circuit* is defined as the circuits between modules and from modules to the common connection points of the dc system.
- *Photovoltaic system voltage* is the dc voltage of any PV source or PV output circuit.
- *Solar cell* is defined as the basic PV device that generates electricity when exposed to light.
- *Solar photovoltaic system* is the total components and subsystems that, in combination, convert solar energy suitable for connection to a utilization load.
- *Subarray* is an electrical subset of a PV array.

There are many similarities between wind-turbine and solar PV installations but also some important differences. Both are used in stand-alone and utility-interactive configurations. Solar power, however, is dc and requires no rectification. For stand-alone applications, the backup battery banks or other energy-storage systems are similar, and both require disconnects and blocking diodes to prevent feedback. Inverters, to create ac for local use or synchronized ac for cogeneration with the utility, are more elaborate in wind systems because of the more rapid fluctuations and high peaking due to gusting winds.

## *NEC* Mandates

*NEC* Section 690, Part II, "Solar System Circuit Requirements," provides information on voltage correction at various ambient temperatures. As noted previously, solar-cell output is greater at lower temperatures, especially when there is no load. And because open-circuit voltage determines the voltage rating of cables, disconnects, overcurrent devices, and so on, the *NEC* provides Table 690.7, which lists voltage correction factors for silicon modules. The temperature range in this table is –40 to 77°F. The table refers to the lowest ambient temperature for the region. Multiply the open-circuit voltage by the correction factor to specify the system components.

*NEC* Section 690.8, "Circuit Sizing and Current," provides information for determining the maximum current in a solar PV circuit. For PV source circuits and output circuits, it is the sum of parallel-module short-circuit currents multiplied by 125 percent. For inverter output circuits, it is the inverter continuous output current rating, taken off the nameplate. For a stand-alone inverter input circuit, it is the stand-alone continuous inverter input current rating when the inverter is producing rated power at the lowest input voltage.

*NEC* Section 690.9, "Overcurrent Protection," refers all questions regarding solar PV system overcurrent protection back to the generic *NEC* Article 240.

*NEC* Section 690.11, "Arc-Fault Circuit Protection (Direct Current)," requires a listed arc-fault circuit interrupter (AFCI), PV type, for in-building PV systems with dc source circuits or dc output circuits operating at or above 80 volts.

*NEC* Part III, "Disconnecting Means," requires disconnects to be placed at all critical locations within the PV system as in a wind system. *NEC* Part IV, "Wiring Methods," resembles requirements for wind systems. PV source and output circuits operating at over 30 volts are to be installed in raceways where the location is readily accessible. Because raceways cannot be attached directly to PV modules, the area may have to be fenced off unless the modules are high enough to be considered not readily accessible. What is decisive is whether a ladder is needed.

Single-conductor cable, not Code compliant in most applications, is permitted in exposed outdoor locations in PV source circuits for modular interconnections within the PV array. For most electrical wiring, the *NEC* prohibits conductors smaller than 14 American Wire Gage (AWG), but 16 and 18 AWG can be used for modular interconnections, provided that ampacity requirements are observed.

PV wiring is not to be installed within 10 inches of roof decking or sheathing unless it is directly below the PV modules or related equipment. It is permissible to run the circuits perpendicular to the roof surface. The purpose of this requirement is to limit risk of shock when firefighters use power saws to cut holes in roofing to vent flames. This hazard is a result of the fact that the PV conductors are not deenergized by a disconnect.

*NEC* Part V, "Grounding," provides that one conductor of a two-wire PV circuit operating at over 50 volts is to be grounded. The same applies to the reference conductor of a three-wire bipolar PV system.

## Cogeneration or Battery Backup?

Both wind-turbine and solar PV power systems can exist as either stand-alone or cogeneration installations. Cogeneration depends on a connection to the utility line, and this depends on it being close by. The electric company invariably figures that a line extension should be built at the customer's expense, and the cost is always higher than expected. This is the rationale for a stand-alone system.

A stand-alone system usually has a battery backup or other type of energy-storage system. Energy storage can take many forms—compressed air, water pumped to a higher level, a railroad car towed upslope, a capacitor bank, or a huge flywheel. The most widely used energy-storage medium is a bank of lead-acid batteries. This is the big problem in a stand-alone system. The batteries, for a bank of any size, are expensive, and their lifespan is finite. Over a period of time, the battery expense may be greater than the cost of a wind or solar installation.

For this reason, utility cogeneration, where a tie to the grid is feasible, is usually the way to go. But this depends on the cooperation of the utility. Because utilities are regulated by a public commission, they often have no choice in the matter. (The operative concept is "kicking and screaming.")

For cogeneration to work, there has to be an electrical service at the building. The wind or solar output is connected to the utility supply. When the wind or solar output exceeds premises demand, the excess is backfed into the grid. Once this is set up, it requires no human intervention and is completely automatic. The utility transformer on the pole, fed in reverse direction, steps up the voltage to

**Figure 14-4**    Reverse metering used with a cogeneration hookup. Notice the disconnect!

match the voltage in the distribution line, and this energy participates in powering all other connected loads. By means of reverse metering, as shown in Figure 14-4, the customer's net contribution of power to the grid is determined, and the utility is obliged to credit this amount to the customer's account and issue a payment if there is a surplus. The amount credited can be figured at the wholesale price of electricity or at the retail rate. This is determined by the regulatory authority and generally is not a decision that is made by the utility.

Wind or solar PV power is dc. With or without energy storage, it can be used to power dc loads. Some loads, such as incandescent lights, do equally well on ac or dc. Home appliances with motors require a standard house current ac, although dc models are available at a premium price. Because most users prefer the convenience of being able to plug in a standard appliance, computer, or power tool, the solar or wind output is usually changed from dc to ac. What is wanted is a well-regulated ac whose voltage corresponds to a sine wave.

The equipment that performs this conversion is known as an *inverter*. The first inverters were rotary machines that can be described as *motor generators*. They resemble electric motors with no output shafts, just closed-end bell housings. A rotary inverter contains dc motor and ac generator windings. Terminals on the outside of the housing allow for connections of the dc input and the ac output. A rotary inverter provides a high-quality sine-wave output, but because it is a mechanical device with moving parts, it is expensive to manufacture and may require maintenance. For this reason, most inverters currently deployed are electronic devices, the sine wave being created by semiconductors. The outputs of the first electronic inverters were crude by today's standards, square-wave approximations of the sine wave. This sort of electricity is good for incandescent lighting and similar resistive loads, but a motor or transformer will protest if fed such a diet by overheating and running at reduced power. Over the years, solid-state inverters have been developed that produce outputs more closely resembling a pure sine wave, and today, there is no problem in powering ac motors.

For the purpose of cogeneration with a utility, a premises with solar PV, wind turbine, or any other alternate power source must feed into the grid electrical energy that is of the same voltage and frequency as the utility power. Moreover, the waveforms must be precisely synchronized. This means that the positive and negative peaks must occur at the same time. The equipment that makes this possible is known as a *synchronous inverter*. By means of internal circuitry, the synchronous inverter continuously samples the utility power at the connection and creates an output that synchronizes with it. This equipment is a factory-made unit that is contained within an enclosure designed for either an indoor or outdoor location. The synchronous inverter is sized according to the output and type (e.g., wind, solar, or other) of the alternate source. It also has the failsafe ability to disconnect the alternate source from the utility supply, in the manner of a transfer switch, in the event of an outage. This is necessary to protect utility workers from backfeed that would energize distribution lines that are being repaired.

Increasingly, wind-turbine and solar PV systems are being sold as packages that include, where desired, synchronous inverters for the purpose of cogeneration. If you are retrofitting a synchronous inverter to an older solar PV or wind-turbine electrical system, you will need to match the characteristics carefully. Beyond this, it is just a question of hooking up the external connections correctly because all terminals will be marked.

Where there is the opportunity to make a cogeneration hookup, a battery backup installation should not be considered because the continuous maintenance cost will be an ongoing liability. Backup battery banks resemble solar arrays in the sense that while you are working around them, you must consider them to be live at all times. The battery bank is to be provided, as previously noted, with a disconnect. When doing battery maintenance, however, this device will not protect the worker because precisely where the maintenance is needed is upstream from the disconnect.

A characteristic of dc is that the electrical connections are prone to corrosion. Oxidation is a reversible electrochemical reaction. In an ac connection, because the polarity is continuously reversing, the oxidation does not accumulate as in a dc connection. Moreover, vented lead-acid batteries release acidic vapors into the surrounding air, and some of this acid finds its way into the individual battery terminations. For these reasons, periodic battery bank maintenance involves taking apart and cleaning battery terminations. In the process, there is the possibility of dropping a metal tool in such a way that it shorts out part or all of the battery-bank circuit. Not limited by branch-circuit conductors, an intense arc-fault can ensue, resulting in personal injury or fire. Such maintenance should not be performed until long after the last charging cycle so that any hydrogen gas will have time to dissipate. A battery bank should be in a dedicated room with ample working space and good ventilation, and this has to be considered as part of the cost of a battery backup system.

Because batteries lose their charge over time as a result of finite internal impedance, they are inherently inefficient. A larger battery bank has greater losses. The battery bank should be sized not too large and not too small. Battery banks are appropriate only for stand-alone applications and not for cogeneration installations. In all cases, lead-acid batteries are the only cost-efficient type. Lithium-ion, nickel-cadmium (NiCad), and other advanced-type batteries are too expensive to consider for backup applications.

Notes on battery-bank maintenance:

- Lead-acid batteries contain an electrolyte that is a mixture of sulfuric acid and water. The greater the charge in the battery, the more acidic is the mixture. Electrolyte levels should be carefully checked and maintained not only because when the level goes down there is less electrical storage capacity but also because if part of a plate is exposed to air and acid fumes, it will rapidly oxidize. This reduces electrical capacity. Also, these areas of the plates will swell, reducing the spaces between them and internal impedance so that the battery will not hold a charge.

- If a lead-acid battery is left in a discharged state over a period of time, the plates acquire a sulfur-compound coating. This is known as *sulfation*, and the battery will not take a charge. Sulfation sometimes can be reversed by applying a dc charge for a period of time. Also, lightly tapping on the outside of a battery case with a rubber mallet sometimes can cause the coating to fall off the plates so that it accumulates harmlessly at the bottom below the plates. Overall, sulfation leads to shortened battery life.

- The amount of charge in a battery cannot be measured directly with a voltmeter. A hydrometer has a rubber squeeze bulb that draws a sample of electrolyte out of the battery so that its specific gravity can be measured. The gauge is calibrated to show percentage of battery charge.

- Ample-sized battery terminals have less tendency to heat up when carrying heavy current—hence there is less corrosion.

- Battery posts and clamp-on terminations should be cleaned with a wire brush. Sandpaper will leave insulating particles embedded in the metal, making for reduced ampacity. Do not scrape terminations with a knife. Corrosion inhibitor can be used to good effect. Sometimes grease is used to keep out acid contamination. This should be applied only to the outside of the joint, not on mating metal surfaces. If it is heated, the grease will dry and make for a resistive joint.

- Once a week, lead-acid batteries should be given an equalization charge. This is about 10 percent higher than the charging voltage, and it should be applied for 8 hours. It makes for longer battery life by intentionally creating gas bubbles, which prevent layering and also mix the electrolyte. If an equalization charge is not applied periodically, there will be a more acidic mix near the top of the battery, resulting in shortened life. An equalization charge is not necessary for the battery in a moving vehicle, where bumps in the road agitate the electrlyte, preventing layering.

At best, lead-acid batteries deteriorate in time and need to be replaced periodically. On the other hand, barring lightning damage or other mishap, an electronic synchronous inverter will last indefinitely, so cogeneration, where possible, is the better choice.

## Foundations for Wind Turbines and Solar PV Arrays

The concrete work for both wind turbines and solar arrays should be of the highest quality, although for different reasons. In the case of a wind turbine, it is a question of resisting the lateral thrust of the highest-velocity wind that will be encountered in the life of the installation. When a three-blade turbine is spinning rapidly, it offers the same resistance to the wind as a solid disk of the same radius. The sideward thrust of the wind will seek to overturn the concrete in the ground or separate the tower from it. Therefore, the tower must be very well anchored, and the concrete must be massive.

For a solar PV array, there is less tendency for the wind to disrupt the array because it is lower to the ground and not facing into the wind. But the issue with a solar PV array is that the concrete work should be absolutely stable so that accurate positioning of the panels is maintained. Moreover, if there is any misalignment, the installation will be unsightly.

The foundation for a wind turbine should be a cube or cylinder of concrete that extends deep into the ground, well below the lowest level frost could ever penetrate with no snow cover. Also, the deeper it goes, the better it will resist the horizontal thrust of the wind at the top of the tower. The top surface of the concrete should be sufficiently above the finish grade that water will not accumulate. It should slope $\frac{1}{8}$ inch per foot for drainage. It should be troweled smooth (but not over troweled) to make a durable weatherproof surface.

Large anchor bolts should be provided so that the tower will never pull off the foundation. They should be electrically bonded to reinforcing rod within the concrete to create a Ufer grounding electrode that is better than ground rods.

For a solar PV array, the concrete foundation is usually formed by means of Sonotubes, heavy cardboard cylinders into which the cement is poured. For individual solar panels to be set at the same level, given the fact that the ground will be uneven, it is necessary for the Sonotubes to be set using a builder's level or transit. The number of Sonotube footings per panel will depend on the size of the panels, the length and diameter of the tubes, the strength of the steel panel supports, and the height above grade of the panels. A solar array with this type of foundation is shown in Figure 14-5.

Sonotube construction has disadvantages, and many installations, where proper procedures are not followed, result in flawed final products. In areas where the ground freezes deep, the concrete footings can be lifted out of the ground, even if they extend far below the frost line. This is so because the ground freezes to the upper part of the concrete cylinder and lifts it out of the ground. This defect can be countered by using a long Sonotube that extends deep into the ground. Another technique is to make tapered footings using lumber or premade forms. In this way, the mass of the concrete is deep, and there is less surface area in the freezing zone. It is also helpful to backfill with a washed, screened, ¾-inch stone. Some builders install a plastic sleeve for the concrete cylinder. This prevents the freezing ground from gripping the footing. Still another technique is, in a separate pour, to make a subfooting for the Sonotube. It should be fairly thick and extend laterally beyond the cylinder. If the two pours are anchored by means of steel bolts, there will be less tendency for the concrete to heave out of the ground.

**Figure 14-5**  Sonotube concrete footings should not be too far out of the ground so as to limit pitting from the weather.

Many finished Sonotube footings exhibit severe surface pitting in the portion above grade. Not only is this unsightly, but it also sets the stage for more deterioration down the road as water penetrates into the concrete and freezes. These *rat holes*, as they are called, are prevented by tamping the fresh cement immediately after it is poured. A long stick is used to poke the cement so that air bubbles rise to the surface and are released. The problem with a Sonotube is that these bubbles rise from the bottom and collect at the top. By the time all the air bubbles are out, the cement has been so overtamped that segregation occurs. This means that the stone aggregate sinks to the bottom, water comes to the top, and nowhere is there a proper mix. Again, the end result is badly pitted concrete work. The solution to this problem is to place the cement in a series of individual 12-inch *lifts* so that the air is released without overtamping. For a durable top surface, it should be sloped slightly to shed water away from the anchor bolts. Then the surface should be smoothed nicely without overtrowelling.

## Fuel Cells

If hydrogen and oxygen atoms are brought close together, they join spontaneously to form water molecules. In the process, for each molecule of water that is formed, an electron is released in addition to a quantity of energy in the form of heat.

A fuel cell takes advantage of this simple reaction to generate electricity. Each fuel cell consists of a cathode, an anode, and an electrolyte. At the anode is a catalyst, frequently platinum. The catalyst facilitates the electrochemical reaction, but it does not take part in it. It is not consumed and does not have to be replenished.

In the most common types of fuel cells, hydrogen is oxidized at the anode. The only waste product is water, so the fuel cell is pollution-free, and there is no contribution to global warming. If dc electricity is passed through water, hydrogen is produced at the cathode and may be compressed and bottled in cylinders or piped to another location. (A useful by-product is oxygen, which is produced at the anode.) If the dc electricity is produced by a solar PV array, there is the potential to obtain storable energy and use it locally or to power mobile vehicles.

# CHAPTER 15

# What Lies Ahead?

The utility meter reader, paying a quick visit once a month to every house on the street, is rapidly disappearing. Instead, a new type of meter has appeared. The smart meter, shown in Figure 15-1, inserts into the existing socket—so as far as installation goes, there is not too much to it. But this device has great hidden functionality, and ultimately, it may affect the way we live our lives because suddenly there is the potential for centralized control of virtually every home in the world.

Electricity has become so much a part of every home and the lives of the people in it! Hand tools are becoming less essential. Even the hammer is powered by air, the compressor having an electric motor. We have telephones that take pictures and cameras that talk, all electrically powered.

There is no force on Earth that can compel you (the home crafter-electrician) to have a smart meter attached to your building, but then the utility can withhold electrical power from the substation, shown in Figure 15-2, from those who do not play according to their rules, as spelled out in the public utilities commission filings. The regulatory bodies are in the business of siding with the suppliers, not end users, of electrical power.

At best, it is a standoff. Time, however, is not on the side of the utilities. This is so because the price of installed photovoltaic (PV) solar systems is in freefall. Solar cells, for the most part, are made of silicon, a very abundant element in the Earth's crust. What accounts for the price of PV solar cells is the great number of operations that are required to create crystalline or thin-film silicon cells that are capable of generating electricity and exporting it into the outer world.

**Figure 15-1**    This smart meter has a two-way wireless connection to the utility.

The stages in the manufacturing process have been simplified and automated so that the cost of solar PV cells in the form of panels, modules, and complete arrays has dropped dramatically, and there is every reason to believe that this trend will continue. A promising developing technology is building-integrated solar PV. Thin-film PV cells may be used to manufacture building-material units that have PV properties as well as comprising the wall and roof coverings. When and if this trend becomes a mature technology, electrical power generation will become decentralized. As things stand currently, utilities will not be out of the picture because cogeneration is still the most efficient means of energy storage. However, a further dynamic is at work. If utilities are compelled to buy surplus premises electricity at retail prices, faced with a diminishing market, they will be able to continue operating only at a substantial loss.

The situation will change abruptly if batteries are developed that are more efficient, longer lasting, and less expensive. Alternative methods for storing energy

**FIGURE 15-2**   The substation is the source of local power, and the utility controls all lines that emanate from it.

include, as we have noted, compressed air, water pumped to a higher elevation, and fuel cells running off of hydrogen that is manufactured by solar-derived electrolysis during daylight hours and stored in cylinders. These and other storage strategies show promise. At present, the impediment is high initial cost. This may drop as manufacturing techniques are refined. The fact is that we are obliged to move in some of these directions in order to reduce carbon emissions and restore the health of our planet's atmosphere.

In my view, the home crafter-electrician can play a key role in all of this. We can build self-sustaining electrical systems that are independent of the grid, and in so doing, we can show our friends and neighbors that there is a better way.

In the past, the word *smart* was not generally attached to inanimate objects, but that is changing. Now there are smart buildings, smart telephones, and smart foods. Although they've been around for a few years, smart meters and the smart grid are gaining importance. In addition to the meter manufacturers and software providers who stand to profit from this implementation, the principal proponents are the utilities. They have a vested interest in expanding their infrastructure in the interest of increased efficiency and an enhanced balance sheet.

## Smart Meter Anatomy

Early (i.e., nonsmart) electric meters worked in a variety of ways. The first arc lamps for street lighting were connected in series, and all went on and off simultaneously, so it was necessary only to clock the hours of usage and compute the electric bill accordingly. Thomas Edison's parallel-connected direct-current (dc) filament lamps were individually switchable, so a more sophisticated solution was required. Edison, always a chemist, built a metering device that consisted of an insulated vessel containing zinc plates hooked in parallel with each customer's premises wiring. Utility workers weighed the electrodes to determine the monthly bill. This method was not very accurate, so it was soon replaced by a motor-driven meter.

With the implementation of George Westinghouse's alternating-current (ac) electrical system, it became apparent that two opposite-phase ac fields could cause an armature to turn. This became the basis for the watt-hour meter that remained in use for over 100 years.

In the 1930s, electrical codes were revised to allow meters to be connected upstream from the main disconnect and overcurrent devices. This, along with socket-mounted sealed units, made customer tampering less likely. For ease of reading, meters were located outside. A gear train with a 1-to-10 ratio conforming to our digital numbering system made meter reading easy once you got past the clockwise/counterclockwise shift between adjacent digits. Since 1990, the four major meter manufacturers (i.e., Landis+Gyre, General Electric, Itron, and Elster) have offered fully electronic models without moving parts. A fully electronic smart meter with wireless connection to the utility is shown in Figure 15-3.

These meters remain widely used today. An important variation is the torroidal coil, a current transformer that senses current flow by means of magnetic flux variations. In heavy-load applications, such as large industrial facilities, this mechanism simplifies wiring and hardware requirements. Overall, the traditional electromechanical meter has worked well for utilities, customers, and electricians. Individual utility policies determine the point of connection, which varies depending on whether the service is aerial or underground. The electrician installs the service conductors and equipment downstream from the point of connection. The utility may or may not supply the meter socket.

For an underground service lateral, typically the customer (or the electrician) furnishes the conduit, usually polyvinyl chloride (PVC), buried, backfilled, and graded. The utility furnishes the wire upstream from the meter because it would not be appropriate to have nonutility workers climbing the pole and connecting to the transformer.

**FIGURE 15-3**  The new smart meter plugs into the existing meter socket, with no additional wiring required.

With the advent of the smart meter, most of these procedures remain unchanged. But depending on its features, a smart meter, for full implementation, implies the use of smart appliances and home automation. Electricians will be the big players in this arena if the homeowners and maintenance departments are unable or unwilling to undertake this work. Moreover, the very fact that home and commercial building owners are becoming more energy aware will mean that they will look for other ways to upgrade their electrical systems in the interest of lowering their electric bills. Electricians can speak knowledgeably about the benefits of good grounding and adequate wire sizing. Poor grounding means out-of-balance legs so that some motors will run below rated voltage, drawing increased current. Moreover, as the years go by, motorized tools and appliances begin to draw more current, producing heat in place of torque. In an industrial facility, significant capital can be conserved by replacing outmoded motors and lighting. The shift to T-8 fluorescent bulbs and ballasts saved large facilities many thousands of dollars annually. On the horizon are light-emitting diodes (LEDs), an enormous improvement in terms of energy use when and if the initial installation cost comes down. In many locations, utilities offer incentives for large customers to improve their power factor by installing capacitors. Power quality is a key consideration in large industrial facilities that are heavily motorized. You can pay for that top-of-the-line oscilloscope by using it to find harmful harmonics that are affecting power consumption and motor life. The smart meter will provide a starting point for this and similar endeavors. These are jobs for the electrician that may result from smart meter installation.

Smart meters work in close to real time to read premises sensors, report power outages, and detect power quality issues. They go way beyond the traditional interval and time-of-use meters that have been around for a quite a few years. An older technology involved the installation of two meters on the outside of a building. One was wired to the entrance panel, whereas the other was wired, through a sealed enclosure containing a 24-hour timer, to the hot-water heater. The idea was that the hot-water heater was powered down during peak usage hours, enabling the utility to distribute its load more evenly. In return for agreeing to this arrangement, the customer was billed at a lower rate for power measured by the second meter. The basic paradigm was that everyone benefited—utility, customer, and our planet—due to reduced carbon usage and less global warming. The problem, however, was that the equipment was more costly than the energy savings warranted, so this dual-meter arrangement has fallen into disuse.

Utilities pay an average of $200 per smart meter, but this figure varies widely depending on buyer, seller, and number of units purchased. For those among us who are interested in such things, it is instructive to take a look at the inner workings of a typical smart meter manufactured by Elster. Needless to say, the unit is not cord-and-plug connected. Two large copper conductors rated at 200 amps are connected to the input and output lugs of a conventional meter socket. Rather than powering an old-world motorized gear train, these conductors are inductively coupled to current transformer windings, whereby a small current is sampled periodically in order to develop the digital information that is transmitted to the utility. Additionally, a small portion of the 240-volt, 60-Hz supply is stepped down to 10 volts, which is subsequently rectified to provide dc bias for the many integrated circuits that populate a printed circuit board. The smart meter has a local-area network (LAN) ID number so that it can be recognized to participate via the Internet Protocol with the EnergyAxis Smart Grid network.

Smart meters probably will gain in functionality as quickly as utility engineers and meter manufacturers can think of new uses. The basic idea is that the smart meter delivers information to both the energy provider and the end user. Both ends of the supply chain, it is stated, will become more efficient, meaning that less energy will be expended to provide more of the benefits we have come to expect. The two parts of the equation are improved decision making by consumers regarding their energy use and more effective management of the grid by providers. The smart grid and smart meter presuppose one another and are codependent.

It was not always this way. In simpler times, the smart grid consisted only of a Supervisory Control And Data Acquisition (SCADA)–enabled information network that aided in maintenance and control of utility-owned substations. This protocol increased efficiency and lessened downtime, but there was no interactivity with customer-owned equipment beyond monthly visits by the meter reader. This is changing rapidly.

Smart meters are similar in appearance to their traditional counterparts, although the curved-glass body that protrudes from the meter socket is replaced by a clear plastic housing with a flat face at the front through which a digital display may be viewed. Internally, there is a radio transmitter, with an internal (or in special cases external) antenna. Typical operation is in the 902- to 928-MHz frequency band in short millisecond pulses. Power consumption is monitored at 15-minute intervals or less, and the information as reported to the utility may be made available to the consumer either online or by means of a control panel installed in the customer's home or place of business. Using this information, the customer can fine-tune appliance usage to minimize consumption during peak periods and reduce the electric bill while helping the utility to even out harmful demand cycles. Moreover, the meter can detect and report outages and aberrations, even functioning as an arc-fault detector for the entire premises.

These benefits are palpable, and everyone stands to gain. The fact is, however, that utilities and manufacturers have been confronted by an enormous amount of opposition on the part of an aroused public, and the answers they have provided have not been satisfactory from the point of view of opponents. Objections fall into several categories:

- **Health.** Long-term exposure to any electromagnetic field (EMF) is seen by smart meter opponents as a health hazard. Numerous studies have supported this point of view, whereas others have countered that the level and duration of exposure are minimal, the danger further mitigated by the fact that the meter is outside. But what about the child who sleeps inside next to the exterior wall just inches from the meter? It seems that for every assertion, there is a counter argument, so we can only say that the jury is out on this one.
- **Privacy.** Two issues are relevant. The smart meter, like some other electrical distribution equipment, is located on the customer's property.

Notwithstanding any legal easements that may exist, bringing in a new technology against the desire of a owner is invasive to say the least. Moreover, this equipment reports activities of the residents to the utility, and these residents have no way to know how this information will be shared. This may not be an issue to you or me, who may have no secrets, but for some, it is an invasive infringement, and the right to privacy for these individuals should not be violated. Additionally, what are the implications of the remotely controllable kill switch?

- **Fire hazard.** It is contended that the smart meter constitutes a palpable fire hazard. This could be true if it is installed incorrectly, but this is true of other electrical equipment as well. But there have been documented instances of this complex instrument overheating and posing a fire risk.
- **Centralized power generation.** Opponents argue that the installation of vast numbers of expensive smart meters further commits us to the idea of centralized power generation, along with a continuing stake in conventional fossil fuel use and associated planetary warming. Moreover, it is asserted that utility rates inevitably will rise to pay for this vast implementation, which is unlikely to be offset by the small effect it will have on peak usage.
- **Vulnerability to cyber attack.** Radio-controlled customer access to the grid, it is argued, makes us increasingly vulnerable to cyber attack. Just as hackers have shrugged off conventional password protection, so hostile political entities or terrorist organizations could exploit vulnerabilities in the North American grid. Smart metering, if it becomes universally adopted, could be a way in. Data hacking, introducing malware into the system, and other attacks at the grid end could shut off the lights.

At one time, the biggest threat was that dishonest customers would attempt to steal power by inverting the old-style meter, deploying strong magnetic fields near the meter, and similar stratagems. Now the danger is that, for whatever reasons, hostile organizations would attempt to compromise the grid so that a large geographic area would be without power, and data transfer would be hampered.

Software providers are attempting to develop security against unauthorized changes in cybernetic configurations in the hope of making the grid more secure and reliable. It is uncertain at this time which side will prevail. The smart meter–grid combination could turn out to be a valid defense against cyber attack or, as a worst-case scenario, a vulnerable point of entry for a destructive assault.

# Glossary

**Ampacity**   The amount of current, measured in amperes, that a conductor can carry without damage, usually to the insulation. The ampacity depends on the size of the wire, the type of insulation, the ambient temperature, the duty cycle, the duration, and other factors.

**Ampere**   A measure of current flowing through a conductor, source, or load. The amount of current is one ampere when $6.25 \times 10^{18}$ electrons per second pass a given point in the circuit.

**Branch circuit**   The conductors between the final overcurrent device protecting the circuit and the outlet.

**Cable**   Conductors usually contained within an outer jacket, as shown in Figure G-1. Additionally, a single conductor that is 6 American Wire Gauge (AWG) and larger without an outer jacket is referred to as cable. A specialized conductor assembly such as submersible-pump cable has no outer jacket.

**Capacitance**   Typically provided by parallel conductive plates separated by a dielectric layer, it is characterized by an ability to store electric charge. It has less impedance to high-frequency electric current and maximum impedance to direct current.

**Circuit breaker**   An overcurrent device that may be reset and reused, as shown in Figure G-2, as opposed to a fuse, which is destroyed by overcurrent.

FIGURE G-1    Various types of cables, raceways, and conductors.

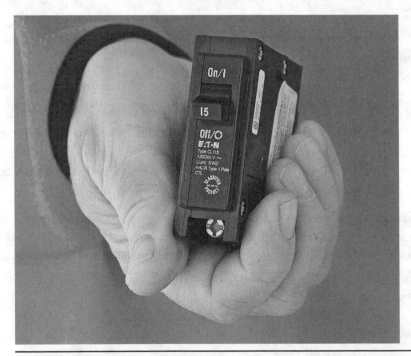

FIGURE G-2    A single-pole circuit breaker, widely used in residential entrance panels.

FIGURE G-3    Schematic diagram symbol for a coil.

**Coil**    Also called an *inductor*, as shown in Figure G-3, a coil is formed when a conductor is wound around a core such as air or soft iron, producing a magnetic field when current flows through it. The coil then exhibits inductance, having greater impedance to higher frequencies and zero impedance (except for incidental resistance) to direct current.

**Conductor**    A material, shown in Figure G-4, usually metal drawn out in the form of a wire, that forms a path for electric current to travel.

**Conduit body**    A fitting in a raceway system, shown in Figure G-5, designed to provide access through a removable cover to the interior of a system to facilitate pulling conductors.

**Continuous duty**    A load that is expected to be connected to the power source for three or more hours.

**Current**    The flow of electrons through a conductor, power source, or load. In a series circuit, where there are no parallel paths, it is the same at every point.

FIGURE G-4    Conductors are available in many colors, aiding in identification.

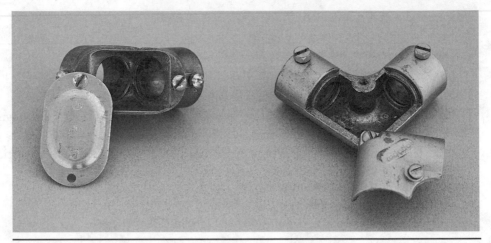

**FIGURE G-5**    The removable cover facilitates pulling of conductors.

**Demand factor**    The ratio of the maximum demand of a system to the total connected load.

**Device**    A unit of an electrical system, shown in Figure G-6, that carries or controls electrical energy as its principal function. It does not consume power except incidentally due to a small amount of resistance.

**FIGURE G-6**    Devices are available in a great many types and sizes for different purposes.

**Diode**   A semiconductor device with two terminals, shown in Figure G-7, allowing the flow of current in one direction only. It is forward biased when positive voltage is present at the anode and reverse biased when positive voltage is present at the cathode. Accordingly, it can be used to change alternating current to pulsating direct current, which can be subsequently filtered to make pure direct current, a process known as *rectification*.

**Equipment-grounding conductor**   A bare or green wire or other conductive material, usually run in cable or raceway with the circuit conductors, that is capable of returning fault current to the electrical system grounded neutral, facilitating operation of the overcurrent device.

**Feeder**   All circuit conductors between the service equipment or other power-supply source and the final branch circuit overcurrent device. A feeder always has overcurrent devices at both ends.

**Fitting**   A locknut, bushing, or other accessory to a wiring system, shown in Figure G-8, that performs a mechanical rather than an electrical function.

**FIGURE G-7**   Schematic diagram symbol for a diode.

**FIGURE G-8**   A great many fittings are needed to make a premises, electrical system work.

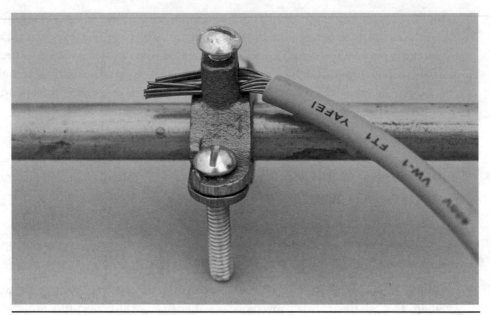

FIGURE G-9    Water pipe as a grounding electrode.

**Grounding electrode**    A device placed in the ground for the purpose of providing a grounding connection to the Earth, shown in Figure G-9. Examples are the familiar ground rod, ground plate, ground ring, and Ufer grounding system inside poured concrete.

**Grounding-electrode conductor**    A wire that connects the grounded neutral, usually inside the meter socket enclosure, to the grounding electrode.

**Inductance**    The property of an electrical device or circuit by which an electromotive force is induced in it as the result of a changing magnetic flux. Inductance may be intentional, as provided by an inductor, or unintentional, for example, parasitic inductance in a transmission line.

**Insulation**    A material, shown in Figure G-10, that restricts the flow of electric current. Electrical conductors are provided with a covering of insulation to protect humans from shock and to prevent fault current from flowing to ground or to other conductors that are at a different voltage potential. The amount and type of insulation must be appropriate to the voltage applied to the conductor and to the maximum anticipated temperature.

**Intersystem bonding termination**    A device, shown in Figure G-11, that provides a means for connecting bonding conductors for communications and other systems to the grounding-electrode system.

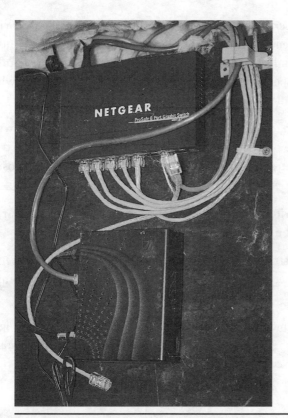

FIGURE **G-10** Insulation is required for most conductors.

FIGURE **G-11** The intersystem bonding termination, in the small white plastic enclosure below the meter socket, is a required part of every new service.

**Main bonding jumper**   A screw, busbar, wire, or other conductor that provides the connection in a service enclosure between the neutral bar and the grounding terminal or metal cabinet and subsequently to all equipment-grounding conductors. The main bonding jumper is field installed because it is not to be used where this connection has been made in another upstream enclosure.

**Main disconnect**   A switch, often taking the form of a circuit breaker, shown in Figure G-12, that is capable of disconnecting premises wiring from the electrical supply. It is located in the entrance panel or within a separate upstream enclosure, indoors or out.

**Molded-case switch**   A switch that may be installed in an electrical enclosure. It resembles a circuit breaker but does not provide overcurrent protection.

**Neutral bar**   A conductive strip provided within an entrance panel or load center, with terminals for the connection of grounded conductors that are part of branch circuits.

**Neutral conductor**   A grounded conductor, color-coded white, shown in Figure G-13, that is connected to the neutral bar in an entrance panel or load center. It provides the return path for current from the connected load.

**Figure G-12**   The main disconnect is frequently located in the service-entrance panel.

**FIGURE G-13**   The grounded conductor is in most systems the neutral conductor. It is white and is not to be confused with the bare or green equipment-grounding conductor, which is at the same potential but serves a different purpose.

**Overcurrent device**   A device capable of providing protection for service, feeder, branch circuits, and equipment over the full range of overcurrents between its rated current and its interrupting rating. Fuses, shown in Figure G-14, and circuit breakers are examples.

**FIGURE G-14**   Cartridge fuses are reliable overcurrent devices.

**Power**   Electrical power is the rate of doing work, measured in watts, and represented by the letter $P$. It is transferred from the source to the load. Watts is the product of amperes and volts.

**Raceway**   An enclosed metallic or nonmetallic channel designed to hold wires, cables, or bus bars. Conduit is one type of raceway. Type EMT is not conduit. It is tubing. Fittings are shown in Figure G-15.

**Reactance**   The opposition of a conductor or load to a change in electric current or voltage due to inductance or capacitance. It is measured in ohms, but unlike resistance, it varies with frequency. Thus there is inductive reactance and capacitive reactance. Reactance and resistance together comprise impedance.

**Receptacle**   A device, shown in Figure G-16, installed at an outlet for the connection of an attachment plug.

**Semiconductor**   A material, frequently crystalline silicon, that is formed of alternate layers of P- and N-type material. The semiconducting junctions will either conduct or not conduct a high-level current as they are variously biased by a low-level voltage. For this reason, they may be used as rectifiers, amplifiers, oscillators, and in digital switching applications. A transistor schematic is shown in Figure G-17.

**Service**   The conductors and equipment, shown in Figure G-18, for delivering electrical energy from the serving utility to the wiring system of the premises served.

**FIGURE G-15**   Type EMT fittings: set-screw type at the left for indoor work and compression-type at the right for outdoor work.

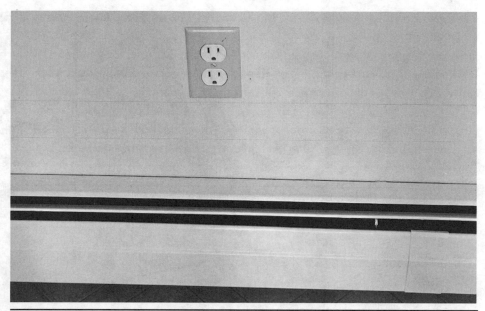

FIGURE G-16   This wall receptacle is improperly located above a baseboard heat unit.

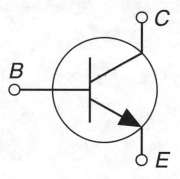

FIGURE G-17   This transistor schematic shows the base, emitter, and collector with terminals attached.

**Service drop**   The overhead conductors between the utility electrical supply system and the service point.

**Switch**   A device, shown in Figure G-19, that is capable of making or breaking an electrical circuit. It can be operated manually or automatically.

**Transfer switch**   An automatic or manual device for transferring one or more load conductor connections from one power source to another. A transfer switch suitable for residential applications is shown in Figure G-20.

**Figure G-18**   Service drop with point of connection and weatherhead.

**Figure G-19**   Rocker switches combined with a ground-fault circuit interrupter (GFCI) are an attractive feature in this upscale bathroom.

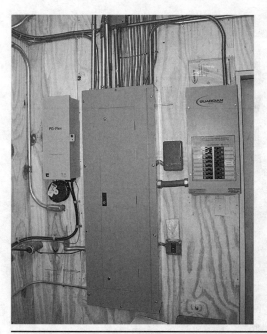

**Figure G-20**    A limited number of selected circuits are chosen to be energized by the alternate power source during a utility outage.

**Transformer**    A device, shown in Figure G-21, consisting of two or more coils wound around the same iron or air core. If an electrical supply is connected to one coil, voltage will appear at the terminals of the other(s). The amount of voltage is determined by the ratio of turns in the primary (input) winding to turns in the secondary (output) winding. Thus it is possible to make a step-up or a step-down transformer. If the number of turns is equal, the input and output voltages are the same. Such a device can serve as an isolation transformer.

**Transistor**    A device that amplifies, oscillates, switches, or otherwise modifies the voltage at two output terminals in accordance with the voltage and/or current at two input terminals. One input terminal and one output terminal are usually common, so the most basic types of transistors have three leads (Figure G-22).

**Uninterruptible power supply (UPS)**    A power supply, shown in Figure G-23, used to provide alternating current to a load for some period of time in the event of a power failure. Unlike a backup power source, there is no time interval without power.

**Utility-interactive inverter**    An inverter, shown in Figure G-24, intended for use in parallel with an electric utility. The device must exactly match voltage and frequency and synchronize the waveform.

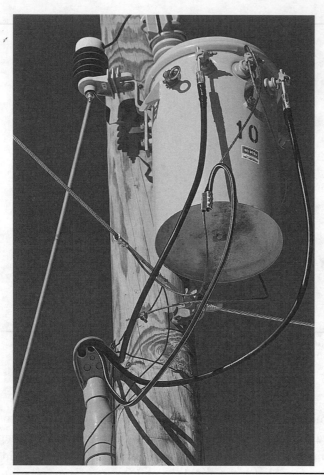

**FIGURE G-21**    A pole-mounted transformer connected to the service for an individual home.

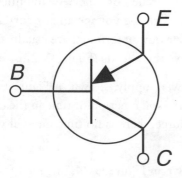

**FIGURE G-22**    Schematic diagram symbol for a PNP transistor.

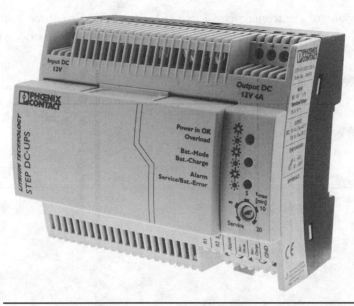

**FIGURE G-23**  A UPS prevents computer crashes when utility power is lost. (*Courtesy of Mouser Electronics.*)

**FIGURE G-24**  A utility-interactive synchronous inverter. (*Courtesy of Kaco.*)

**Volt**   The amount of electromagnetic force (EMF) required to force a current measuring one ampere to flow through a load having an impedance of one ohm.

**Watt**   A measure of the amount of power transferred from a source to load. Dc amperes times volts is equal to watts.

# Index

*References to figures are in italics.*